GROW YOUR OWN ENERGY

Edited by
MICHAEL CROSS

Basil Blackwell & New Scientist

First published in book form in 1984 by
Basil Blackwell Limited
108 Cowley Road, Oxford OX4 1JF.

Basil Blackwell Inc.
432 Park Avenue South, Suite 1505
New York, NY 10016, USA.

British Library Cataloguing in Publication Data

Grow your own energy.
1. Biomass energy
I. Cross, Michael
662'.6   TP360

ISBN 0-85520-731-0
ISBN 0-85520-730-2 Pbk

Typeset by Oxford Verbatim Limited
Printed and bound in Great Britain by
Bell and Bain Ltd, Glasgow

Contents

Contents

Among the contributors

ANIL AGARWAL is director of the Centre for Science and the Environment, New Delhi. He is *New Scientist*'s correspondent in India.

MICHAEL ALLABY is a freelance writer; *JIM LOVELOCK FRS* is visiting professor of cybernetics at the University of Reading, England.

NIGEL CALDER was editor of *New Scientist* from 1962 to 1966.

GERRY FOLEY AND GEOFF BARNARD are the authors of *Biomass Gasification for Developing Countries*, published by Earthscan, London.

DAVID HALL is professor of biology at Kings College, London. His work and publications have contributed greatly to the understanding of biomass energy.

SIR HAROLD HARTLEY FRS capped a long career in chemical and energy engineering by becoming chairman of the Energy Commission of the Organisation for European Economic Cooperation in 1955. He died in 1972.

MICHAEL JONES is lecturer in plant physiology at the Botany Department, Trinity College, Dublin.

TREVOR LONES is a freelance journalist who lives in Brazil.

LIONEL MILGROM is *New Scientist*'s chemistry consultant.

PETER MOORE is senior lecturer at the Department of Plant Sciences, King's College, London.

KEITH OPENSHAW studied wood consumption in Tanzania, Thailand and Gambia for the Food and Agriculture Organisation of the United Nations.

STEVE PRENTIS is staff editor of *Trends in Biochemical Sciences*.

AMULYA REDDY is secretary of the Karnataka State Council for Science and Technology, India.

ERIC SENIOR works at the Department of Bioscience and Biotechnology, Applied Microbiology Division, University of Strathclyde, Scotland.

MALCOLM SLESSOR is director of the Energy Studies Unit, University of Strathclyde, Scotland.

BILL WOLVERTON AND REBECCA MCDONALD are researchers with the National Aeronautics and Space Administration's National Space Technology Laboratories in St Louis, Mississippi.

A note on units

New Scientist follows the SI system of units as far as possible. However some other units are occasionally useful when talking about energy. One that crops up is the British Thermal Unit (Btu) measure of energy:

1 Btu = 1055 joules
1 therm = 1055 megajoules

Another non-metric unit is the barrel of oil. This is normally about 35 gallons, or 147 litres.

Billion = 1000 million, throughout.

Commercial quantities of energy, such as electricity, are usually measured in watts. The watt, of course, is a unit of consumption, equivalent to 1 joule per second.

Some crops are measured in bushels. One bushel is about 35 litres.

Foreword

New Scientist was born on 22 November, 1956 and, with only a few interruptions, has appeared every week since. It was intended as a "popular" magazine about science, and so it has remained; but as one of that elite among popular magazines that also serves the specialists. It has brought together all who are concerned with what science has to say about the world: scientists, technologists, politicians, industrialists, teachers, and a new breed of science writers, some of whom are journalists who found that the ideas of science were the most exciting of all, and some of whom are ex-scientists, who found that commentating was a fine and proper way to indulge their interest.

New Scientist Guides are collations of articles taken from the magazine, sometimes abbreviated a little to avoid repetition, but otherwise presented as they first appeared. The individual pieces have been brought up to date and put in perspective by linking passages, to give a multi-faceted, modern overview of the subject in hand.

Thus we intend the Guides to fulfil two functions. First they describe the main lines of thought and the key events in a given area of science or technology in the past few, crucial decades; but they also show how those events were brought about, how the ideas were first framed, and how they were first perceived by the world at large.

By presenting science and technology in this way the Guides illustrate what is perhaps the greatest scientific discovery of all: that the traditional view of science as an inexorable progression to unequivocal truth is misguided, and that science is, in fact and in essence a product of human thought and human imagination. Articles appear in *New Scientist* in various guises. Some – roughly distinguishable as those over 1000 words – are acknowledged to be "features", and are printed with the names of their authors: so they appear in the Guides, together with their date of publication.

Shorter pieces tend to be placed in one or other of the magazine's special sections, variously titled "This Week", "Monitor", "Technology", "Comment", "Forum" – and, in the old days, "Notes and Comments". Most of these short articles are published anonymously, which again is how they appear in the Guides, together with their dates and the name of the section in which they appeared.

Linking passages, printed in sans-serif, have been written specially for the Guides by the Guide editors, and did not appear first in the magazine.

Colin Tudge
Series Editor

Introduction:
Why grow your own energy?

When people talk about an age of solar energy they are usually referring to some undefined period in the future. But for most of its history the human race has depended entirely on the Sun, exploited in increasingly sophisticated ways, for that useful commodity we call energy. The Sun is a massive resource: about 3×10^{18} megajoules of energy from it fall to the Earth's surface every year. This is something like 20 000 times as much energy as the human race consumes. However, the Sun's energy arrives in a fickle and dispersed form. Most of the history of energy technology has consisted of attempts to concentrate the Sun's power into a more compact package and to even out the effects of cloud, night and winter.

The further energy technology moves from the primary resource, the more versatile and reliable it becomes. Spread your crops on the ground for the Sun to dry them, and they are at the mercy of the weather. But if you build a turbine to catch water in a mountain stream and generate electricity from it, you have converted solar energy (which of course hauled the water up the gradient in the first place) into a dependable and versatile resource. Dry your crops by day or night, so long as the stream keeps flowing, and watch television at the same time.

But for all the engineers' skill at concentrating, processing and storing solar energy, nature has done a far better job with the mechanism of photosynthesis. A field of corn, a forest, even a stretch of scrubland, are all stores of concentrated solar energy. To exploit it at its simplest level, you simply gather up the "biomass", dry it if necessary, and set fire to it. This technology has served the human race from the dawn of history until the present day. And as we shall see, it shows no sign of dying out.

What eclipsed simple biomass as the Western world's main source of energy was the discovery that nature had provided solar energy in even more concentrated forms, the fossil fuels. It is important to remember how sudden has been the rise of fossil fuel, and what a versatile source of energy it is. The most important fossil fuel, coal, came into widespread use in Europe and the US only in the 19th century. In Britain, thanks to the deforestation caused largely by the population boom of the 18th century, the change came earlier. But animal, wind and water power still fuelled most of Britain's industrial revolution – the world's first.

Coal is a remarkable material. It is compact, containing roughly five times as much burnable material as the equivalent weight of wood, and there is a lot of it about, though the resource seems to be concentrated in comparatively few places. Apart from the pollution it causes, coal is an excellent fuel for the giant thermal power stations that supply the developed world with its electricity. In the heady days of the 1960s, as *New Scientist* reported at the time, electricity seemed the answer to the world's energy problems. Vast national grids, driven either by coal or by nuclear power, would span the world.

The trouble with electricity, however it is generated, is that there is no practical way to store it efficiently. This is as true today as it was in 1956. Motor vehicles, aircraft and ships need a fuel that is light, compact and potent. This is where oil comes in. Oil and natural gas have risen in prominence even more recently than coal: they overtook it as the most important commercial sources of energy only in 1960. But 13 years later, the "oil shock" of 1973/74 reminded the world that its supplies of oil were limited and unevenly distributed. Although the oil price rises of the 1970s were not in themselves the result of scarcity (indeed by making fields such as Alaska and the North Sea economically viable, they actually increased estimates of the world's recoverable resources), the oil shock sparked a new debate on the wisdom of recklessly consuming finite fossil fuels. At the same time came the realisation as development policies ran into stagnation that most of the world's people, particularly in the agricultural countries of the Third World, would never taste the "good life" of cheap oil.

One answer has been to investigate the challenge of new sources of energy: nuclear, solar photovoltaics, fusion perhaps. There has even been talk of putting giant arrays of solar cells into orbit, to catch the full resource of the Sun. But none of these new sources has yet proved a practical proposition. The only one that comes close is nuclear power, which few countries have the money or infrastructure to exploit.

Another promising area of research is to process coal to make it cleaner and more portable. "Coal-liquefaction" schemes flourished in the wake of the oil shock, but apart from special cases such as in South Africa, they cannot yet compete with oil. In Britain, work has already begun on producing "synthetic natural gas" from coal for the day that Britain's natural gas deposits run out. In countries such as Britain with adequate supplies, coal should have a great future, not least as a feedstock for many chemical processes. But the resource is again finite, and unevenly distributed: the US, USSR and China have around 85 per cent of the world's hard coal between them. Countries without coal will have to look elsewhere.

The obvious thing to do is to conserve energy. Thanks partly to the recession, partly to new attitudes, the developed world has already become rather less profligate in its consumption of oil, at least. The western world has burnt less oil every year now since 1979. But the West's concept of conservation – turning down the central heating, driving at 80 kilometres per hour instead of 120 km/h – are irrelevant to much of the Third World.

The rise in oil prices hit developing countries as much or more as it hit the West. Some countries, such as Tanzania and Turkey, found themselves spending more on oil than they earned from exports. People in the villages may not have noticed the effects of rising oil prices as dramatically as the American motorist did, the long-term effects to them were more severe. During the 1970s many a development project died as the governments of the Third World were caught between the low price of the commodities they exported and the high price of imported manufactured goods and energy.

Coupled with the essentially political and economic story of the "energy crisis" came the realisation by scientists that most people in the world were going to be stuck with biomass energy for the foreseeable future. It is not a pleasant outlook. To an affluent ecologist in Europe a wood-burning stove might appear a benign technology when compared to the ugliness of a coal mine or the sealed menace of an atomic power station. But in much of the world the reality looks very different. In countries where people depend entirely on wood for fuel, the picture is often one of a constant scramble for survival, in which the exhausting daily round of collecting wood leaves little time for productive tasks. Such a life-style can be ecologically disastrous – when land is stripped of trees for fuel the topsoil soon disappears, bringing famine and misery. When people do not have wood, they will burn dried animal dung as food, robbing the soil of valuable fertiliser. The end result is the same as removing trees.

The challenge facing science is to develop biomass fuels that enhance, rather than erode, the environment, and which improve the quality of life in the Third World – and eventually the developed world, too. That is what this book is about.

There are many ways to classify biomass technologies. One is to separate those that require processes such as fermentation from those that involve simply harvesting a crop and burning it. Another distinction is between strategies based on waste products and those that grow crops specifically for fuel. There is some overlap; many but not all waste products need processing, while many fuel crops do.

Broadly speaking, this guide organises its subjects in order of the complexity of the technologies involved. We begin with a look at how the energy picture has changed over the past 20 years, and move into the potential of simple biomass crops and the problems of woodfuel in the developed and developing world. Slightly further up the scale, we turn to the production of biogas from fermented wastes, then into the "high-technology" world of distilling fuel alcohol from crops. Finally, we take a tentative look at research that may open the doors to a new energy technology, harnessing photosynthesis to split water or even to generate electricity.

Even if such possibilities do not turn out – and the history of energy technology is littered with blind alleys – science can do a great deal to improve the efficiency of biomass energy.

No-one has yet come up with a practical biomass system that can offer the convenience of hydrocarbon fuels such as coal and oil. Indeed experience suggests that to develop renewable energy we need to sacrifice some of the attractions of conventional energy sources.

Only time will tell if this is true. What does seem certain is that a biomass fuel that can step into the shoes of the hydrocarbons will not be a "soft" technology. Such a fuel would require an industry on the scale of today's oil companies, but producing a product that everyone, not just a privileged minority, can afford. The product will have to be non-polluting and compact, able to power cooking stoves, motor vehicles and generating plants. Most importantly, such a fuel will have to be truly renewable, and cultivating it must not deplete the soil or take land from food crops.

In many ways this goal is as far off today as it was when the first issue of *New Scientist* appeared. In those 28 years we have learned much about introducing new technologies that should make us more cautious. But on the credit side, we know a good deal more about the problems and possibilities of biomass energy, and our first taste of "energy crisis" – albeit an artificial one – has sharpened our sense of urgency. We now know that sooner or later the whole world, not just the poor one, will have to grow its own energy.

PART ONE

Changing ideas

New Scientist was born into a world with an energy crisis: the oil embargo that followed the Israeli, British and French attack on Egypt in autumn 1956. But it was also a world of technological optimism. In October 1956 electricity from Calder Hall, the world's first nuclear power station, entered Britain's National Grid. In its first issue, a month later, *New Scientist* reported that "the novelty of generating electricity by nuclear power has already worn off". In December that year it reported that "The whole energy picture has been changed by the future possibilities of atomic power . . . atomic power stations may be built wherever cooling water and the necessary technical resources are available."

Technological confidence was everywhere. Articles from the 1960s suggested that fuel cells and magneto-hydrodynamic generation would supply cheap and clean electricity to the world. The developing countries would follow the same course to prosperity as the Western world had done.

It's sometimes easy to smile at the presumptions of that age, which are exemplified here by Sir Harold Hartley's article on "fuel and power in 1984". But looking at them 20 years on, it is striking how little has changed. Hartley estimated that by 1984 the world's energy consumption would be about 10 billion tonnes of coal equivalent a year. He was about right, although his estimate that liquid fuels would supply more than solid ones was out, and he greatly over-estimated the impact of nuclear power on the total – as opposed to the electrical – energy scene.

Hartley's article on 1984 was part of a long series that set out to look at a world 20 years hence. A rather touching "get-out 'clause' " introduced the series: "It is recognised that a certain amount of guesswork is unavoidable and that unforseeable discoveries and inventions could, in some cases, radically alter the picture." In fact no new discoveries or inventions were to alter the energy picture radically. What changes that came followed the political and economic phenomenon of the oil shocks of the 1970s. Chapter 3 "Technological alternatives", dates from this time, and sums up the changes in thinking that came with the 1970s. Big technology was out and "alternatives" were in.

Nowhere was the change in thinking more apparent than in the Third World which, strangled by colonial underdevelopment, never had access to cheap energy anyway. Two articles on India (Chapters 4 and 5) show how perceptions of the energy picture have changed over the years. The first, from 1962, looked forward to an approach based on a centralised "hard technology" with electrification and nuclear power hauling the country into an industrial age. The second, published in 1981, examines the consequences of that strategy in a country that has to import its oil, and urges a future based on alternatives.

By 1981 alternatives had become respectable enough to warrant the attention of an international conference under the auspices of the United Nations. We conclude with reports from the conference. It was clear that, far from Hartley's prediction that non-commercial energy would be of "declining significance" for most people it was more important than ever.

1

Fuel and power in 1984

SIR HAROLD HARTLEY, FRS
28 May, 1964

This article, part of a series of predictions on the world in 1984, was ahead of its time in describing the crucial relationship between energy and development. While some of Hartley's technological optimism now seems misplaced, his assessment of the "energy gap" is still relevant today.

The policies of every nation, developed or emergent, are directed today toward economic growth. This depends on the availability of energy, the life-blood of industry and the vital factor in domestic comfort. So energy must play an important part in any economic forecast for 1984.

Figure 1 (p. 4) shows the role of energy in a nation's well-being. It plots the average per capita income of 50 countries against their consumption of energy per capita. The coordination between the two is partly causal and partly due to effect. Abundant energy is needed for prosperity, and this in turn raises the standard of living and the consumption of energy. Other factors such as climate come in to complicate the picture and no accurate statistics are available for the consumption of non-commercial sources of energy (wood, vegetable waste and dung) which still play an important part in the more backward countries.

However the overall relationship between energy and well-being is abundantly clear. How much energy is the world likely to consume in 1984? It will depend on the average consumption per capita and the growth of population. If the present trend continues the average consumption of commercial energy per capita would be about 2.25 tonnes of coal equivalent, and with a projected world population of 4500 million, the global energy consumption would be 10 billion tonnes of coal equivalent a year. The figures neglect the

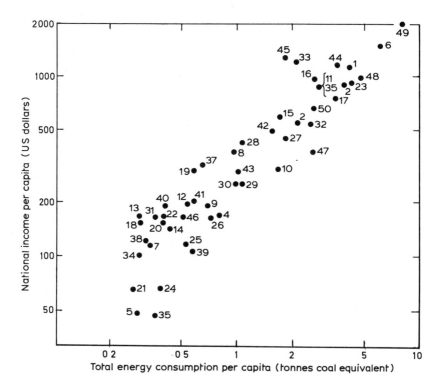

Figure 1. *Comparison of per capita national income in US dollars and per capita energy consumption from both commercial and non-commercial sources for selected countries (1958).*

1. Australia	17. West Germany	34. Nigeria
2. Austria	18. Ghana	35. Norway
3. Belgium and	19. Greece	36. Pakistan
Luxembourg	20. Guatemala	37. Panama
4. Brazil	21. Haiti	38. Paraguay
5. Burma	22. Honduras	39. Peru
6. Canada	23. Iceland	40. Philippines
7. Ceylon	24. India	41. Portugal
8. Chile	25. Iran	42. Puerto Rico
9. Colombia	26. Iraq	43. Spain
10. Cuba	27. Ireland	44. Sweden
11. Denmark	28. Italy	45. Switzerland
12. Dominican Rep.	29. Japan	46. Turkey
13. Ecuador	30. Mexico	47. South Africa
14. Egypt	31. Morocco	48. United Kingdom
15. Finland	32. Netherlands	49. United States
16. France	33. New Zealand	50. Venezuela

non-commercial sources, which are of declining significance. In 1961 the world consumed 4600 million tonnes of coal equivalent, taking into account the amount of coal saved by hydroelectricity.

It is not so easy to assess the proportions in which the amount of energy will be provided by the various primary sources. Predictions have to take into account the trend of consumption, relative price trends and the economic progress of nuclear power and possibly of magneto-hydrodynamic (MHD) generation.

The major change in the energy pattern in this century has been the increasing demand, both from industry and householders, for energy that has been upgraded into a more sophisticated form such as electricity, tailored liquid fuels, gas and smokeless fuels that are easy to distribute, handle and use.

In 1900 probably 90 per cent of energy was consumed in the raw state as wood or coal, and by 1929 that had fallen to 50 per cent. By 1961 about 85 per cent of fuel was upgraded. Today, about a quarter of the input of primary energy is consumed as electricity, the electric motor having displaced the steam engine in industry and an infinity of uses having been found for electricity as the most sophisticated form of energy. The refining of oil has provided the liquid fuels that are basis of modern transport in nearly all its forms, and they have many other uses. The same is true of natural gas.

The global forecast for 1984 has also to take into account the different trends of consumption of the three main consumer groups of developed countries (North America, Western Europe, Communist) with the populations, consumptions and the growth rates. In 1961, 49 per cent of the world's population consumed 86 per cent of its energy. If the consumption of electricity continues to increase at 7 per cent compound, it will rise to $10\ 000 \times 10^9$ kW/h in 1984. Supposing that 10 per cent is generated in nuclear stations in that year, their contributions would represent about 400×10^6 tonnes coal equivalent. A significant contribution from MHD by 1984 is problematical.

Taking all these factors into consideration, the changes in the consumption of primary energy from 1961 to 1984 may look something like Figure 2 (p. 6). These estimates will serve at any rate as a basis for discussion, and forecasts will undoubtedly differ. By the 1980s we shall know which is right; with the rapid pace of technological development some quite unexpected changes may occur. The estimate for solid fuels takes into account the rapid rise of consumption in the Communist countries, which now consume 50 per cent of the world production of solid fuel with an annual

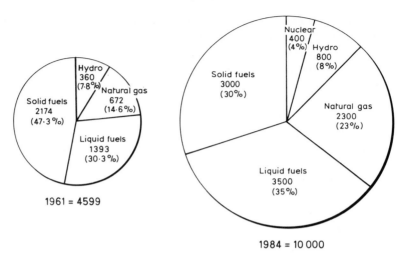

Figure 2. *World (commercial) energy consumption in 1961 and 1984 (in millions of tonnes of coal equivalent).*

increase of more than 12 per cent since 1951. A large increase in the consumption of coal in North America is also forecast during the next 20 years.

These global figures, and the adequacy of world reserves to meet them, conceal the energy problems of individual countries. The pattern of energy consumption depends partly on the availability of indigenous energy sources. Among the advanced countries, the USSR is exceptional in being a net exporter of energy. Britain, Europe and Japan are particularly dependent on imports of oil and even the US is a net importer. Moreover, the sources of energy are very unequally distributed between nations, as the following list shows:

Location of 90 per cent of proved and estimated resources
Coal: US, USSR, China
Oil: Middle East, US, USSR
Natural gas: North America, Middle East, USSR, North Africa, Netherlands

Comprehensive estimates of water power potentials are lacking; they appear to be more equally distributed although here again the US and USSR have the major resources. The advanced countries are able to make good their deficiencies by imports, for which they can afford to pay, but even there the flow is distorted by duties, quotas and taxation.

Much thought is being given today to the problem of the emergent countries and to technical aid. How do they fare as regards energy? Figure 1 shows the low level of consumption per capita, and even where the gross consumption is increasing their rapid growth of population tends to nullify this advantage. With the exception of a few countries with large oil resources, notably in the Middle East and Africa, the developing countries lack indigenous resources of energy adequate to their population and their area, because energy is costly to transport over long distances except by tankers or by oil or natural-gas pipelines. By and large, energy presents no insuperable problems for the further development of the advanced countries, on whose progress in the future so much depends. The transport of energy in the form of nuclear fuel is the cheapest of all methods, so nuclear stations may in time provide for part of the needs of countries with inadequate indigenous energy resources.

A considerable effort has been directed to the development of the so-called "new" sources of energy for use by the emergent countries. These include geothermal heat, wind power and solar radiation. Geothermal heat is available only in a few localities; wind power is intermittent and it is expensive to store the electricity it generates; solar radiation on the other hand has considerable possibilities for domestic use in simple appliances and will undoubtedly help by 1984.

The emergent countries, with their dense populations living in small towns and villages, need energy badly for light, for village industry, for the irrigation of crops and drainage and for the local processing of harvests. Energy for transport is also essential for their development. The solution to the developing world's energy problems should be one of the first objectives of technical aid if the gap between the developed and the emergent countries is to be narrowed by 1984.

2

The world's changing energy pattern

WILLIAM C. DAVIDSON

29 August, 1968

In 1968 the oil shock was still 5 years away and high technology still held its promise. But that year's meeting of the World Energy Conference, in Moscow, gave an airing to some alternative technologies and encouraged scientists to look hard at the environmental consequences of burning fossil fuels.

The world's reserves of energy are now generally agreed to be sufficient for as long as anyone can foresee. But the fact remains that they are badly, and unfairly, distributed demographically. Thus energy transportation, both within and between countries, has to be resorted to, to obtain a better spread of natural wealth.

This emerged as a recurrent theme last week in Moscow, where the World Power Conference held its seventh plenary meeting since its foundation in London in 1924. Before getting down to discussing the 270-odd papers, however, the 2500 delegates heard that the name had been changed to the World Energy Conference, to bring the English title into line with the French and Russian equivalents, and also to remove any possible misunderstandings about the nature of the organisation.

Raw energy – coal, oil, gas, water and uranium ore – is found wherever nature happens to have left it. Moreover, it is quite useless in the as-found state: it must be refined, transported and converted into heat or mechanical power. Electricity is one form of energy transportation, and it too has ultimately to be converted into usable forms. At the moment, a surprisingly small part of the world's energy is converted into electricity. In the United States, for example, probably the most electricised country in the world, electricity accounts for less than 30 per cent of the nation's gross national energy consumption (although this should reach 50 per cent by the turn of the century).

Already most West European countries are interlinked electrically as a form of mutual aid, and the Eastern bloc is also building up its so-called "Mir" network, by means of which all those countries will be hooked into the Soviet supply system.

There is still no single system for the whole of the USSR, but Soviet engineers plan to achieve this by about 1980. Big plans are afoot to develop the almost uninhabited – and one would have thought, uninhabitable – TransBaikal region, and once this is done the local time difference between there and Moscow (7000 km away) should simplify peak-load problems.

Technically there is no reason why such ideas should not be applied between different countries on a global scale, but this would clearly imply continental states, if not a world state.

Transporting electricity is, however, not always the answer, for it can cost quite a lot: if the fuel is of high calorific value it can prove cheaper to transport the corresponding electric power. In the US, for example, the projected Mohave Power station on the Colorado River will be fed with coal brought as a water slurry through a 430 km pipeline from a mine in northern Arizona.

On the subject of pipelines, the Russians have a project that seems to beat anything so far. Vast natural gas reserves have been discovered in northern Siberia near where the River Ob flows into the Arctic Ocean. To bring this gas to western Russia 3000 km away, a pipeline no less than 2.5 metres in diameter is to be laid. This will be able to carry 100 000 million cubic metres of gas a year – 10 times as much as is carried by a normal present-day Russian pipeline and at 40 per cent less cost per cubic metre.

On the question of coal it is interesting to note that world production is steadily increasing, whatever may be happening in Britain and to a lesser extent in France and Germany. The USSR and the US show steady increases every year.

Nuclear energy as we know it is now regarded as practically conventional, but there are quite serious doubts about economic uranium reserves over the next decade owing to the very poor uranium utilisation of the thermal-neutron (as opposed to fast) reactor. Consequently there is now a unanimous call for all efforts to speed the advent of the fast breeder.

Steam turbines continue to grow in size (the largest ordered so far is for 1300 MW, by a US company) but the growth rate is definitely levelling off. Two main factors account for this; the size of the low pressure end and the generator leading to transport problems, and the fact that as far as a power system is concerned one can have just too many eggs in one big basket.

Although of high efficiency, the steam turbine still leaves a lot to be desired, and various ideas were discussed for reaching the magic efficiency figure of 50 per cent – a sort of "four-minute-mile" of the turbine world. These included magneto-hydrodynamic "toppers", ammonia or freon "bottomers" to replace the awkward low-pressure steam end, and hybrid cycles using gas turbines as well as steam.

A sizeable part of the conference was taken up with hydro power – a peculiarly satisfying form of energy, often positively contributing to the natural environment, and self-renewing as long as it keeps raining somewhere. The most spectacular of all current projects is probably the Krasnoyarsk power station on the Yenisei River in Siberia. With three 500 MW water-turbine/generator sets already installed and nine more to follow, this will be the most powerful power station in the world – 6000 MW. And the Soviet plans to build four more stations on the Yenisei, altogether poducing 25 000 MW from this one river.

On tidal energy, three British engineers came up with an idea which is a development of France's great plant near Dinard. Having yet another look at the perennial Bristol Channel, they concluded that a new scheme on these lines could provide 11 700 GWh of peak-time energy a year. The Solway Firth was also examined in this way and the authors urged an investigation of these schemes.

An American delegate foresaw thermonuclear fusion – now also virtually abandoned by Britain – just creeping into the picture before 2000. No progress was reported on this, but come it surely must. Speaking, not at the Moscow conference, but at one IAEA specialist session at Novosibirsk only 2 weeks previously. Henry Seligman, ex-Harwell and now one of the IAEA's deputy director-generals, described fusion as "one of the most important scientific questions of our time, perhaps in all history".

For the rest of this century, in advanced countries anyway, it looks like this: an optimised national nuclear fuel cycle involving both sodium-cooled fast breeder reactors and advanced thermal converters burning uranium, plutonium and maybe thorium; pumped storage almost anywhere it is required, using Sweden's great contribution to the conference; fossil-fuelled power stations getting better, but probably not very much larger, and with much better control of environmental pollution.

But who knows? A real breakthrough on thermonuclear fusion could destroy this picture practically overnight.

3

Technological alternatives

ANDREW MACKILLOP

22 November, 1973

Of course the energy crisis, albeit a political rather than an absolute one, appeared before any breakthroughs in energy technology. Within weeks of the end of the 1973 Middle East war *New Scientist* reported, in its introduction to this feature, that: "The changing energy picture may force us to alter our approach to energy technology. Centralised utilities with massive plants and complex distribution networks may have to give way to decentralised alternatives that conserve resources and do not pollute the environment."

It is very hard to define alternative technology, but its fuzzy outline masks a wide range of non-traditional technology research and development that stretches from applied solar energy, wind power, and microbial systems, through to experiments in social and cultural organisation. This sets many people against the idea of "alternative" technology, because it appears to be radicalism in a new technological guise.

Despite this, or possibly *because* of it, more and more individuals, institutions, and groups are pressing ahead with investigations into non-traditional technologies. Passing over the cultural overtones we come to probably the most explicit theme amongst all those groups involved in alternative technology. This is a basic environmentalist theme, with an emphasis on energy-efficiency, resource recycling, decentralisation, and low pollution.

For the community-oriented alternative technology groups, that stretch from the Biotechnic Research and Development (BRAD) group in Wales through to very large, almost town-like groups in New Mexico and Australia, the development of new energy sources and recycling systems is not the primary aim. Such ecologically-based communities, which encompass as many as 2500 people in more than 10 countries, are seeking new social, political, and cultural organisation, within a theme of environmental acceptability. This does not mean that the techniques they use are in any way unsuccessful. The Findhorn community in Scotland, for example, obtains fantastic crop yields, but uses methods that would make an "agribusinessman" weep.

Despite the fact that such communities do not stress the development of new technology, interesting and effective processes and equipment are being developed within them. A good example is the New Alchemy Institute, with three centres in North America, which is already well-known for its aquaculture work, and has now developed effective methane-producing plant for small farms. Several US communities also have constructed innovative and effective solar-heated domes, and because of the unregimented nature of US planning laws (which usually produces a proliferation of urban sprawl) these are spreading very rapidly, particularly on the West Coast.

Moving away from ecologically-based communities we come to work being conducted by institutions and special organisations that relates to alternative technology. The well-known organisation Intermediate Technology Development Group (ITDG) is continuing its broadly agricultural work, but now has a well briefed power panel, and under Professor Dunn at Reading University is looking into that perenially-remembered low-pollution engine, the Stirling-cycle unit.

The Rolls-Royce Anti-Poverty Group, based at the RR Technical College, has developed a low-cost direct-focus water heater, now being tested in Nigeria; and has plans for several other power-related projects. Yet other groups, from the German International Development Foundation, Volunteers for International Technical Assistance (VITA) and the Swiss HYDRA group, through to work in association with the UN Environment Programme all have explicit or implicit reference to the development of new energy sources.

The Canadian government is also keen to sponsor work in alternative technology research, and plans are slowly shaping-up for a small-medium size township to be built and serviced, in Canada, by ecologically-based processes. Several other national governments are also studying development proposals which contain references to the use of solar energy, domestic waste recycling, very high levels of home insulation for energy conservation, and related processes.

B

Big corporations, for so long wedded to high technology – notably space technology spinoff (such as better hedge trimmers and motorised pogo sticks) – are now moving into R & D that dovetails with many themes in alternative technology. The Arthur D. Little Corporation, in its introduction to a study on solar energy for housing, acknowledges that solar-electrical conversion is both high technology and a long way off. For this reason, it continues, the quickest pay-off in the energy conservation and environment protection fields will come from application of solar energy for household space heating and solar-absorption cooling.

This conclusion, which interestingly point research into household use of solar energy in the direction of low-cost flat plate collectors and bulk heat storage systems, is amplified by an unexpected source, the RAND Corporation. RAND has unequivocally concluded that there are no short or medium term solutions to the energy shortages now afflicting California in particular, and industrial nations in general. It therefore concludes that there should be application of such proven technology as roof-mounted flat plate collectors for domestic water heating.

While the international development foundations and the big corporations tend to leave aside the more innovative work in new technology, this is proceeding very rapidly in spite of a shortage of capital. A host of small and medium-sized companies in the US and Europe, often working in association with disenchanted high-technology engineers and scientists, are looking into such novel concepts as solar-source heat pumps, wave generators, and other items.

In the US there are now several firms selling wind turbines and conducting their own research into design and performance; others are working on such unusual items as hydrogen-fuelled cars; and recurrently "forgotten" ideas, like wood gas, and ethanol fuel from fermentation of hydrolysed waste paper, are cropping up.

Domes are often thought of as somehow primitive structures, akin to their distant forebears the Kazak Yurt, or even the Anglo-Saxon mud hut, and opponents will use the one rational complaint – they are hard to subdivide internally – as the climax to their argument. This throws up another significant theme in the background to much alternative technology thinking and work. This is the "soft" technology argument.

Western society has a crisis of materials and energy supply, and more than a couple of environmental problems resulting from the way energy and materials are used. Virtually all preindustrial societies are organised in extended families, large vibrant groups of people living closely together. The "nuclear" family is part, possibly a major cause, of our "soft" crisis. Removing the intimate barriers inside the home, as the domebuilders have found, is very risky in the short term, but it can pay off handsomely in the longer term in deeper social relations.

The home could become a focus for the early use of non-traditional technologies and renewable energy sources. Research work on ecologically-conscious building and servicing is now progressing at many universities in the US, Canada, France, Germany, and the UK. In Britain several university architecture departments are looking into the use of solar heating, aerogenerator windmills, and methane generation, for example. But it is in the *integrated* use of these non-traditional systems that the most exciting concepts are emerging.

At Cambridge University's architecture department, staff and students are now in their second year of "autonomous servicing" research. Using conventional economic analysis, James Thring has pinpointed the thresholds for systematic use of locally-produced power and waste handling. Briefly, he has found that homes and dwelling groups more than about 50 km from large towns and cities are, on *economic* grounds, candidates for the integrated use of renewable energy sources and domestic aerobic waste handlers such as the Swedish "Clivus".

Two California University scientists, C. Golueke and W. Oswald, have recently published details of their "algal regenerative system" which, when applied to a small home incorporating a livestock shed, offers tight recycling loops and requires very little in the way of fossil fuel energy subsidy.

This exciting concept, now being tested at pilot level on the university's field station, derives most inorganic energy from methane, with waste gas from combustion and sunlight being used by algae in a roof-top tank to give animal feed, flushing water, and soil conditioner. All unused food, animal wastes, and human faeces and urine flushed from the WC with dilute algae pass through the methane-producing tank, situated at the home's core to ensure that it is buffered from external temperature variation. Such is the designed-for level of self-sufficiency that the home uses a solar still to purify water, and in addition has a rain catchment tank. Obviously such tight recycling and the proximity of animals would make this design anathema to our concrete desert dwellers, but such an elegant and practical – and potentially economic – housing design could be of great use in the housing programmes of developing nations. However, would-be popularisers of such an idea should tread warily lest the Third World rejects it on the grounds of the easy inference that it rates humans no higher than animals!

This brings us to a final point, the social and political – as well as economic – implications of the development of alternative technology. There is now a well-rounded collection of case histories showing the negative human and economic results flowing from misjudged prestige technology projects.

But decentralised, small-scale technology does little for central governments, and also decentralises political power. This is as true in the West as anywhere, and must be a factor in careful deflection of proposed developments that seek to undo centralised systems. While the alternative technology movement – which is a model of decentralisation – already has a coterie of self-elected spokesmen, the hard political factors operating against alternative technology

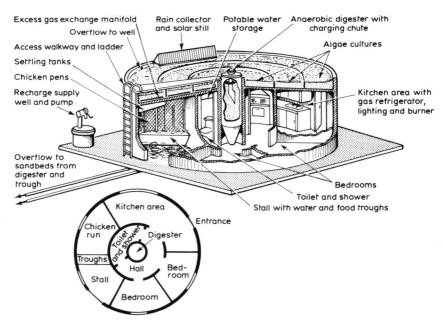

New styles of living are an integral part of the philosophy of alternative technology. This diagram shows "a dwelling unit for a family of four and their livestock which incorporates a micro-biological recycle system for water, nutrients and energy in a convenient and hygienic environment".

are strong and intractable. But cracks are appearing. Individual nations, such as Tanzania, already use development processes that mesh well with localised small-scale systems, and large funding bodies like the World Bank and the Peace Corps are turning more-receptive ears to such proposals.

When alternative technology becomes "respectable", and attracts the support that it eventually must, we can be certain that the present slow and small-scale groundwork will not have been in vain. To this end, factors as apparently diverse as the environment and energy "crises" will provide very useful incentives to the un-plugging of barriers preventing greater use of non-traditional, decentralised, and small-scale technologies.

4

Fuel and power for the development of India

NIGEL CALDER

18 October, 1962

While the rich world played with alternative technologies, the poor world lived with them. Two articles of India (this one and chapter 5), published nearly 20 years apart, illustrate vividly how attitudes in the developing world changed after the oil shock.

In Bihar some months ago a leading Indian metallurgist remarked to me that it was twice as difficult to accomplish anything in India as in the industrially advanced countries. The Indian industrial scientist, hampered by the lack of foreign exchange and the prevailing inadequacies of indigenous industry, has to make in his laboratory many things, whether plant or instruments, which in the normal way would be ordered from outside; he has to train the craftsmen to make and use them; and when his work reaches a stage at which, in an advanced country, development workers within the industry would take over, he has to continue for years in pilot plant work to convince the industrialists and administrators that an Indian scientist knows what he is talking about. The great achievements in India in the past 10 years remain the activities of relatively few dedicated men trying to awaken a vast agrarian country from the slumber of centuries.

When the Indian government announced the award of the contract for its first nuclear power station, which is to be built at the fishing village of Tarapur on the coast 96 km north of Bombay, it recalled to my mind the sight of Indian labourers at the Atomic Energy Establishment, Trombay, working with simple tools and carrying baskets on their heads, just as their ancestors did when they built the great Moghul fortresses. Unless one understands how the old and the new India have to serve one another, and the contrast between the modern industrial cities and the 500 000 remote villages where little has changed in a thousand years, one cannot grasp the nature of the technical and human problems facing those who seek to reinforce the weary muscles of a vast nation with modern resources of fuel and power.

Energy means much to the Indians: modern towns and factories (better distributed over the country), more efficient cottage industry, more economical railways and, not least, light in the home – light to study by, light to extend the days cut short by sunset. It means, too, the hope of saving 300 million tonnes of cow dung which is burnt on domestic fires each year, and using it to fertilise the soil, and of preventing the alarming deforestation and soil erosion brought about by the consumption of firewood. Moreover, it involves the development of heavy industry. For example, at Bhopal a big factory is being developed by Heavy Electricals Ltd, in collaboration with AEI of Britain, for the production of turbines, alternators, transformers and switchgear. When the foundation stone was laid four years ago the intention was to aim at an annual production of about £5 million worth of equipment by 1970; by the time the first products (transformers) were completed last year the 1970 target for the Bhopal works had increased eightfold – to £40 million.

Although the Tarapur nuclear power station and great hydro-electric projects like the Bhakra Nangal scheme (with a dam 200 metres high across the Sutlej river), have more glamour, coal remains one of India's most important and paradoxical problems. There are vast resources of coal but it is coal very different from what we know in Western Europe. The great coalfields of East India were formed along the river valleys and then mashed up by glaciation, so that the coal is of inferior quality mixed up with sedimentary matter which leaves a great deal of ash and also makes the

Pattern of domestic fuel usage in India (million tonnes coal-equivalent)

Fuel	Urban	Rural	Total
Coal	2.0	–	2.0
Electricity	0.2	–	0.2
Kerosene	0.3	–	0.3
Dung	4.0	35.0	39.0
Firewood and farm wastes	13.5	42.0	55.5
Total	20.0	77.0	97.0

coal quite unsuitable for coking. Only about 1 per cent of India's
coal is of coking grade (perhaps 1000 million tonnes recoverable)
and this has important consequences for the Indian steel industry.
The steelworks which were the pride of the Second Five Year Plan
(1956–61) encountered acute difficulties on this score; and at the
National Metallurgical Laboratory at Jamshedpur, one of the most
important industrial experiments in India is the pilot low-shaft
furnace aimed at the production of pig-iron using carbonised lumps
of briquettes in lieu of coke. The director, Dr B. R. Nijhawan, is
convinced that plant of up to 200 000 tonnes capacity can now be
set up in regions of India which, because of the poor quality of the
coal, iron ore and limestone available, were not previously regarded
as potential steel areas.

For coal used as a source of energy, one of the problems is to
design boilers which will cope with coal yielding perhaps 50 per cent
ash – and then using the ash constructively, for example in cement
and fertilisers. But for rural India the big problem is the production
of smokeless fuel, if cow dung and firewood have to be replaced
(and at present only 2 per cent of domestic fuel is coal) the fuel
scientist has to remember that in the villages there are no proper
hearths and chimneys. At the Central Fuel Research Institute at
Jealgora great emphasis is being laid on low temperature carbonisa-
tion of coal, to produce a soft coke for domestic purposes and a
chemically rich gas as well. It has a "semi-commercial" plant which
is a modified version of the plant developed at the Fuel Research
Station in Britain. As the director, Dr A. Lahiri, remarked, the worst
thing you can do with coal is to burn it. "The coal in Assam is a
chemist's dream" – and Dr Lahiri's dream is that it should be used as
the basis for a chemical industry analogous to the petrochemical
industries in the advanced countries.

The coal problem cannot be separated from that of transport. A
large part of India's best coal is used by the railways and the
shocking inefficiency of steam locomotives is now recognised: here
electrification has an important part to play. Moreover, the Indian
transport system is not suited to the transport of large amounts of
coal over great distances – especially the low-grade coals which
make up by far the greatest part of the fossil fuel available for Indian
consumption. Accordingly, industry based on coal is bound to be
concentrated near the coalfields, in the general area to the west of
the neck of the Bay of Bengal.

Next to coal, hydroelectric generation is India's most important
source of energy. The hydroelectric potential of India is about

41 000 MW, although about half of this is in remote mountainous areas where the demand for power is small. Nevertheless, it is envisaged that by 1976 half the estimated installed generating capacity in India will be in the form of hydroelectric plant. It is worth noting that a few years ago the electricity requirements were being greatly underestimated and the latest plans call for very rapid increase. Hydroelectricity is, of course, closely linked with irrigation work. Because of the remoteness of some of the sites long-distance power transmission lines are required: the Third Five Year Plan (1961–66) calls for an increase in high-voltage circuit miles from 84 000 to 150 000.

Other possible sources of energy, which could be important for local use, are wind power, solar energy, tidal power and geothermal power, but they are not likely to make an important contribution to the total energy supply in the foreseeable future.

The gap, if any, between resources and needs will have to be filled by nuclear energy. There have been misgivings about whether the very considerable scientific effort which India has contributed to nuclear energy research and development might not have been better used for more humdrum purposes.

Nevertheless, there is good reason to think that, in 20 years or so, conventional means will not be sufficient to provide India with the power she will then need. Moreover, because a nuclear power station is virtually free from fuel transport problems, it will be possible to site nuclear power stations where coal and waterfalls are missing. And, what is particularly important, India has vast reserves of thorium in her widespread monazite sand deposit – a potential nuclear fuel, in the sense that it can be converted in a nuclear reactor into fissionable uranium-233. But to carry out this conversion on a sufficient scale, rich nuclear fuel is required to start the "breeding" cycle; and the necessary skill and experience in operating nuclear plant have to be built up over a period of a good many years. These are the reasons why India is investing in nuclear activity at this relatively early stage in her development. At the Atomic Energy Establishment at Trombay there are three research reactors working: Apsara (1956) designed in India; the Canada India reactor (1960), a joint Indo-Canadian project carried out under the Colombo Plan; and Zerlina (1961). The first is moderated by ordinary water and the latter two by heavy water; Zerlina is intended to test reactor concepts using a variety of liquid moderators.

The decision to include the construction of an actual power

station in the Third Five Year Plan was taken in August 1958, and Tarapur was chosen as the site nearly three years ago. Tenders were invited from all over the world and now General Electric in the United States have won the contract. It is a first step which will, by itself, make only a small contribution to India's increasing power supplies, but may prove to be a wise investment for the future. At the least, it is a symbol of the determination of the small body of men who hold the destiny of 400 million people in their hands to heave their country into the 20th century by an effort of will.

5

India – a good life without oil

AMULYA REDDY

9 July, 1981

India, like all developing countries that import oil, is heading for a catastrophe because of the massive rises in oil prices. Unfortunately, most of the measures India's leaders have proposed to tackle this crisis have been based on a narrow sectoral approach. They ignore the interdependence of energy-consuming sectors; for example transport and households both rely on oil, but transport can sustain higher prices than domestic users can. I believe that the resolution of the oil crisis lies in an alternative, two-pronged strategy involving measures to be taken in the household sector.

India's oil consumption has been steadily increasing – the average annual compound rate of growth has been about 7 per cent since 1975–76. Indigenous production accounts only for about one-third of the country's total consumption which was about 33 million tonnes in 1980. As two-thirds of the country's needs are imported, the country's oil import bill has been escalating rapidly and has reached about 80 per cent of the estimated total foreign exchange earnings in 1980–81. This already impossible situation can only worsen. It is, therefore, a matter of the country's survival that it formulates and implements a strategy for resolving India's oil crisis.

About 60 per cent of the country's oil is used in transport and 20 per cent in homes. The transport sector uses petrol, aviation fuel and diesel – in 1978 these accounted for about 12, 9 and 79 per cent respectively. Trucks, buses, railway engines and ships all burn diesel, but trucks (64 per cent) and buses (22 per cent) consume about 86 per cent of the total diesel used.

Diesel locomotives haul more than a third of the country's freight while consuming about 11 per cent of the diesel. In contrast trucks carry 36 per cent of the freight, while using six times as much diesel. This illustrates the well-established energy efficiency of railways compared with trucks. Despite this, the share of the total freight carried by railways has steadily decreased.

Trucks have raised their share of freight mainly by increasing the average distance over which they operate. The costs of both truck and rail transport (in terms of resources that move any commodity) increase almost linearly with distance. But while truck costs are lower than rail's over short distances, they rise more rapidly and exceed those of rail at large distances. In other words, there is a break-even point below which trucks use less energy and above which rail transport is more economical. The break-even figure decreases as the diesel price increases – but, on the basis of a 50 per cent increase over 1979 diesel prices, it is between about 100 and 130 km depending upon the commodity. But in practice, the average distance over which trucks are presently carrying freight is three to four times the break-even figure.

The main reason for this proliferation and expansion of truck transport – apart from factors such as shortage of wagons, delays, and inefficiency, which are not unavoidable features of railways – is that truck operators are subsidised. The price of diesel in India is relatively low: only about a third to a half of the price of petrol.

An apparently obvious device to shift freight traffic from road to rail is to increase the price of diesel so that it is on a par with the price of petrol. But this is not possible because the price of diesel is pegged to the price of kerosene. This fuel is produced by the same refineries that turn out diesel, and it can be used in diesel engines. So, if diesel prices are increased while those for kerosene are not, truck operators switch to kerosene – as has happened recently. But kerosene prices cannot be raised without causing great hardship to the poor because, at present, kerosene is the fuel most households use.

Out of this household consumption of kerosene, about 70 per cent goes for lighting and the rest for cooking. More than 87 per cent of rural households and 38 per cent of urban households, together constituting three-quarters of the country's 116 million households, depend wholly on kerosene for lighting. Indians burn a great deal of kerosene – about 4 million tonnes in 1978–79 compared with about 10 million tonnes of diesel – and the country imports about 20 per cent of this fuel. But this is a poor deal for the country because kerosene lamps are very inefficient. They produce about 200 times less light per unit of energy than even the electric light bulb.

Though electric lighting will dramatically improve the quality of life in India, there is alarmingly little electrification of homes to date.

All the 2700 towns (with a population of above 10 000), but only 44 per cent of the 567 000 villages have been electrified. Only about 16 per cent of India's electricity goes to rural areas. And most – 87 per cent – of this rural electricity goes to agriculture to drive pumps, with only 13 per cent being used by households. Rural electrification has mainly served the pumps of the more affluent farmers. Only 14 per cent of households in electrified villages have electricity. And while 1 million households per year are connected to the electricity supply, the number of new households grows by about 2.2 million each year. In other words the number of un-electrified homes is increasing although fewer villages remain without electricity

The first prong of an alternative strategy is to provide all homes with electric lighting. Kerosene will then become unnecessary for lighting, removing the constraint on diesel prices which can then be increased, for instance, to the level of petrol prices. If, in addition, a permit system for trucks and buses is designed (and enforced) to make these vehicles operate within the distances at which they can operate economically, there will be a drastic reduction in oil consumption.

The National Transport Policy Committee has shown that, using realistic estimates of freight traffic, the projected diesel consumption for the year 2000 decreases from about 30 million tonnes to about 14 million tonnes by shifting to rail 75 per cent of the inter-regional traffic that trucks now carry beyond their break-even distances (on the basis of oil at $30/barrel). But, oil prices may go up to $50 or even $100/barrel in the next few years, implying that break-even distances will be even lower. Also much more inter-regional freight traffic – say 95 per cent – can be shifted to rail and so, offer still greater reduction in oil consumption. Thus, it seems possible to bring oil consumption to below India's indigenous production, about 13 million tonnes/year.

Even this large reduction in oil consumption may not be enough, for two reasons. One is that intra-regional or short-haul traffic is best served by road transport. The other is that road transport may have to shift to renewable energy sources (alternative fuels) if it is to have any future. Hence, the necessity of a second prong to the strategy.

Of the various renewable energy fuels, vehicles run most easily on those derived from biomass – producer gas, methanol, biogas and ethanol. Ethanol – the Brazilian choice to counter the oil problem – is produced from crops such as sugar cane that require good agricultural land, leading to serious competition with food production. Hence, ethanol is unsatisfactory for a country with a high population density such as India. Biogas cannot be easily liquefied and is, therefore, suitable only for transport over short distances. The only feasible alternative fuels are producer gas and methanol (which is obtained from producer gas). In fact, producer gas generated *in situ* with charcoal gasifiers was used in buses and trucks throughout India during the Second World War.

But both producer gas and methanol need wood, and the wood crisis in India is just as serious as the oil crisis. At present, about 130 million tonnes of firewood per year is consumed for cooking, and firewood is becoming more and more difficult to produce. Any attempt to use wood for producer gas and/or methanol in trucks will create even greater shortages of firewood and aggravate the problems of the poor for whom it is the main cooking fuel.

Firewood, however, is not an efficient cooking fuel – the efficiency of firewood stoves (chulas) is about 5–10 per cent. More efficient cooking fuels will not only provide greater convenience to housewives, but also conserve firewood. Fortunately, India has a large number of cattle – the country's human–cattle ratio is about $2:3$. The large amount of cattle waste can generate biogas through anaerobic fermentation, and this biogas can be supplied to every rural home to meet its cooking needs (about 0.87 m³/household/ day). The cooking needs of urban homes – which are now being met by firewood, kerosene, soft coke and liquefied petroleum gas – in 73 per cent, 17 per cent, 5 per cent and 5 per cent of homes respectively – can be met by piped mixtures of sewage, coal and producer gases. If gaseous cooking fuel is thus provided to all homes, not only will the 30 per cent of the kerosene presently used for cooking become unnecessary, but the 130 million tonnes of firewood being used for cooking will become available for other purposes, and in particular, as a renewable biomass fuel.

Out of this 130 million tonnes of firewood, about 75 million tonnes is enough to replace completely (after considering efficiency losses) the 10 million tonnes of diesel that trucks and buses burn every year.

Further, water-lifting pumps can be run on producer gas or methanol. The 4.4 million diesel pumps projected for the year 2000 (the present number is 2.7 million) would consume only about 16 million tonnes of wood instead of 3.5 million tonnes of diesel. And if no further electrical pumps are connected beyond the present 3.6 million, and the expected increase of 7.4 million pumps by the year 2000 are run on producer gas, about 27 million tonnes of wood is required. In other words, 43 million tonnes of wood is sufficient to replace all the diesel that would be used in diesel pumps as well as the 17 000 MWh of electricity that would be required to run the 7.4 million extra pumps. And the 130 million tonnes of firewood released by providing sophisticated gaseous cooking fuels to all homes is more than the 118 million tonnes of wood needed for all diesel trucks and buses as well as all the diesel pumps and additional electric pumps. And the law of diminishing returns can be turned on its head. Once the pressure on trees as a source of cooking fuel decreases, the conditions for growing energy forests will improve drastically, and much more firewood will be available.

Finally, the 17 000 MWh of electricity saved by not installing more electric pumps is only a little less than the 21 000 extra MWh that would electrify all the unelectrified homes in the country. The net increase in electricity consumption by electrifying all homes but not adding more electric pumps would be 4000 MWh in 2000. But it is not essential that the extra electricity for home electrification should come from grids supplied by centralised generation; decentralised generation from local sources such as biogas powered generators and small hydroelectric plants can make a major contribution.

The two-pronged strategy described above shows that by providing electric lighting and efficient gaseous fuel to all homes, India could move toward a virtually oil-free road transport system along with a dramatic improvement in the quality of life for its people. In fact, it appears that the country has been engulfed by a grave oil crisis because it has ignored two crucial basic needs of households: efficient sources of energy for lighting and cooking.

6

More light than heat
6 August and 20 August, 1981

The United Nations Conference on New and Renewable Sources of
Energy (abbreviated, inevitably, to UNERG) saw biomass come of age.
Delegates at the meeting in Nairobi and demonstrators outside called
attention to the fact that biomass was not a declining anachronism
but the world's biggest energy problem. But could a UN talking shop
help? The following three articles are reports from the conference.

The only crisis is firewood

Kenyan women marched through Nairobi's streets last week, carrying bundles of firewood and holding up placards saying "Trees for
people" and "Day-long journey for fuel – relieve us of back-breaking
labour". The destination of the marchers' short walk was the
Kenyatta Conference Centre, where they were received by Kurt
Waldheim, secretary-general of the UN, Prime Ministers Pierre
Trudeau of Canada, Thorbjorn Falldin of Sweden, and Edward
Seaga of Jamaica. They were glad to see the marchers, the dignitaries
said, as it had helped them to understand the severity of the crisis.

The benefit of the march is less clear. Most observers saw it as a
media event, while in the working committees, the major groups
struggled over "forms of words" that would not commit them to
action stronger than they had already planned.

The severity of the firewood (known as fuelwood in the jargon)
crisis is unquestionable. Millions of people rely on wood for cooking, heating and lighting. But how do you do anything about it?
Trees take time to grow. One answer is more efficient cooking
stoves. But then what do you do about the heat and the light?

Any journey through the Kenyan countryside at night confirms
the triangle of needs. In the townships, the larger shops and houses
are lit by primus lamps burning pressurised kerosene, while in the

A carnival atmosphere greeted delegates to the United Nations Conference on renewable energy – but carnivals are not noted for deciding on action.

bush the villagers sit round fires until bedtime. Where you see these fires, there does not seem to be any shortage of trees, but the statistics say differently: in Kenya people are using up the living firewood resource six times faster than it is being renewed.

Wood accounts for over 74 per cent of the country's total energy consumption and the rural population is increasing rapidly. In 1978, the Ministry of Natural Resources estimated that the demand for firewood is rising at up to 2.5 per cent per year.

Kenya is perhaps not a good example of the severity of the crisis, because it is rich by comparison with the least developed countries. But it is still in dire straits. Philip Leakey, a junior minister in the Ministry of Environment and Natural Resources, told a session of the non-governmental organisations' (NGO) forum that Kenya is in a position of survival. Roughly 35 per cent of the country's foreign exchange pays for oil, to service a small proportion of the population. Kenya will have to continue importing at least this percentage of oil for the foreseeable future and there is little chance of per capita income rising.

Leakey says that Kenya needs to "respond fast". But there are few options open. Of those even fewer are practical. So the government is looking at those renewable energy sources that might be helpful. The country has little wind, plenty of Sun, not much hydropower, but lots of arid land for biomass. So biomass it seems to be, largely because it needs little new technology, says Leakey.

But before a country can act, it must know how much of this resource it has. Kenya, says Leakey, has assessed this. The Kenya Range Environment Monitoring Unit (KREMU), originally set up to count wildlife, has prepared a report using pictures from the Earth-resources satellite Landsat, aerial photographs and ground studies. The report is still confidential but according to Leakey, it is encouraging enough for his ministry to start tackling the problem of how to increase the use of biomass from the present level of 2.4 per cent of the country's area.

Another speaker at the NGO meeting took a different line. Philip O'Keefe, of the Beijer Institute of Sweden, which helped to produce the KREMU report, says that increasing conventional forestry as recommended in the report is not the answer – most firewood is collected from outside the forests or consists of cutting down young saplings growing in forest areas.

The result is that most of Kenya's forests consist of mature trees, with nothing else coming after, posing another problem; how to restore a healthy balance in the forests.

O'Keefe does not believe that wood is really short in Kenya, but the labour to collect it is. And this comes back to the backbreaking work that the women on the fuelwood march complained of. The wood available is so far from the people who need it that it is either impractical or uneconomic (in both energy and cash terms) to collect it. So the answer, says O'Keefe, is to plant wood crops outside the forests.

But here again there is an important constraint. You can plant fast-growing trees outside the forests, if you have the labour. Nowadays, when plenty of children go to school, the burden falls on the adult population, and they are busy planting crops for food at the most suitable time for planting trees. So trees tend to be planted after the crops and consequently do not do so well. That is a constraint, says O'Keefe, that is difficult to overcome. And that sums up the problem of firewood.

Biomass may not need much new technology but it does require money that countries such as Kenya do not have. And it also means that governments have to persuade rural people that they have traditionally gathered wood for free. Leakey, who asserts that the present order must change, offers no explanation of how hisgoing to do that. Many people at this conference are waiting for answers to questions like this.

Renewable energies "as bad as nuclear war"

New and renewable resources of energy may well be vital to developing countries which have no conventional fuels of their own but they offer no immediate panacea. They will be costly in terms of money, manpower and energy itself to produce. But the greatest risk of widespread use of new and renewables is to the physical resources of developing countries, the forests, the water-tables, and the soil. And, according to Dr Maurice Strong, former secretary general of the United Nations Environment Programme (UNEP), their indiscriminate use could be as destructive as a nuclear war.

Strong was addressing a panel discussion organised by UNEP. Earlier speakers, including the energy ministers of Kenya, Jamaica and Bangladesh had discussed the energy problems of developing countries, but none had come to grips with the real effects on the environment. Most admitted that there were some bad effects, but the general consensus was that all in all they were minor compared with those of oil, coal, gas and nuclear power. Not until Strong had his say did any note of constructive pessimism creep into the panel's views. Strong's line was that energy had "made it" as a topic of world interest with most countries now boasting ministers specialising in it. And most understand that "renewables" are expensive in economic, social and environmental ways.

These ministers face very difficult decisions, he said. They will be tempted to ignore the costs of opting for the opportunity to cut their oil bills, by exploiting what renewables they have.

For many such energy technologies it is difficult to perceive what the dangers will be. But, said Strong, environmental considerations must be in at the beginning of any new energy schemes. It will be hard enough, he said, to change the habits of engineers, administrators and politicians, but it will be harder still to do it if the effects on the environment are left till some time later.

The prime dangers come from the very energy source that offers most potential – biomass, encompassing woodfuel, biogas, alcohol production. Biomass energy systems, if not carefully implemented, could increase the rates at which forests are depleted, speeding up desertification, altering the water tables and competing with food crops for limited financial and technical resources. As with the technical potential of new and renewable energy, the problem is one of lack of adequate information.

But what then can developing countries do to reduce their dependence on oil? Not a lot, it seems, if they have to place environmental considerations high on the list. Basil Buck, Jamaica's energy minister, explained why. His country now gets 99 per cent of its energy from imported oils. Its largest earner of foreign exchange is its best bauxite and alumina industries, but they account for 50 per cent of the country's energy consumption. In every other area of the island's life, oil-based energy is high on the list. Jamaica has one of the highest per capita uses of oil in the world – 7 barrels per year of fuel oil equivalent. It has few conventional resources.

So, says Buck, the case for renewables in Jamaica is strong. And his ministry has produced an energy policy that is designed to cut its oil consumption by nearly 60 per cent by the year 2000. In the short term it will concentrate on finding conventional energy resources as well as implementing conservation measures. It will try to diversify its "energy mix" by investigating coal and peat as fuels, while the long-term goal is to go "hell for leather" into new and renewables.

This programme could, Buck admits, create environmental problems for future generations, so the Jamaican government will work closely with UNEP, under the Caribbean Action Plan, to minimise these problems.

Moving toward a new energy mix – can renewables really help?

New and renewable sources of energy now contribute about 15 per cent of the world's total energy supply and, according to the UN working party that has been laying the ground for the Nairobi conference, this could rise to about 25 per cent. But the "synthesis group", as the working party is called, also reckons that if developing countries are to develop even modestly, they will need to increase their consumption of energy threefold in the same period. And, if renewable energy's contribution is to nearly double in effect, then the total capacity of the new energy systems will have to rise by five or six times.

The synthesis group probably does not intend to understand the case when it says that "the implications of this challenge are far reaching, involving structural changes in the world economy ...". It goes on to say that if it does happen, "it has profound implications for the rural sector where even a relatively small per capita increment (in energy supplies) will result in a major improvement in quality of life in terms of improved health, education and income for those billions who will be directly affected". The message is clear – it will require a massive effort to increase the energy from new and renewable sources, but every small gain will have a disproportionate benefit.

What are these new and renewable sources of energy and to what extent are they now available? First in importance is solar power, because it is so widely available in developing countries. It can be used both as a source of heat and for generating electricity. But solar energy is diffuse, meaning that it is expensive to collect on a sufficiently large scale to compete with conventional energy sources. Advanced solar cells, producing electricity directly, still cost far too much for general use even in the West, although the technology is advancing quickly.

Developing nations will probably benefit more in the short term from biomass. In simple terms this means anything that derives from photosynthesis. In practice it covers high energy plants, wastes such as forestry cuttings, dung and wood. Biomass can produce liquid fuels, be used for large- or small-scale generation of electricity, and can provide heat. It is already the most widely used form of renewable energy in the Third World.

Fuelwood and charcoal are also widely-used renewable energies in the Third World. Firewood is so widely used that supplies are now at a premium. In some rural communities, such as in Tanzania, the search for wood occupies most of the day for some villagers, during which they can walk up to 50 km from home. The Food and Agriculture Organisation of the UN says that 90 million rural people are already short of firewood and that 800 million people are lopping branches off trees faster than the trees are growing.

Hydropower, large and small, is one of the world's most innocuous energy sources. It is also plentiful. Purely in hydro capacity, the world could replace almost all its present fossil and nuclear stations with hydropower (given a few rather difficult political deals) says Daniel Deudny in a report by the Worldwatch Institute. But more important perhaps is that hydropower has most scope for expansion in the Third World. Asia, Africa and South America have exploited 9 per cent, 8 per cent, and 5 per cent respectively of their potential; in contrast the figures for Europe and North America are 59 per cent and 36 per cent respectively. There are suitable sites for both large-scale plants and mini-hydro stations to provide a total of 1400 GW of energy. All the Third World needs is a little help – money, training and technology – from its friends.

Many people already use tiny windmills for pumping water and could be persuaded of the benefits of going for bigger, more modern machines that could generate heat or electricity. But it is expensive to build a modern windmill that is both efficient and strong enough to stand up to gales. And the energy is often generated when it cannot be used, and lacking when it is needed. Developments in energy storage are awaited eagerly by all proponents of periodic energy sources.

The oceans contain masses of energy – in tides, waves in thermal gradients and there is no shortage of ideas to convert them into a form suitable for use by man. But tidal power and wave power both need storage to give the best return for what is inevitably a massive investment. They are no small-scale applications. The same applies to using the sea's thermal gradients in ocean thermal energy conversion (OTEC). They work by using the differences in temperature between the surface and the bottom of the sea to drive a classic heat engine; this produces steam to power turbogenerators, or which can power processing plants to extract hydrogen and ammonia from the sea. OTEC plants need pipes, 1000 metres or more long that will stand up to bad weather, and must be kept on station in the areas of best thermal gradient. This is a high technology energy source that daunts even the developed nations.

The term new energy sources normally refers to some advanced forms of traditional resources – like peat or biomass, or to oil shale and tar sands. Peat, for example, is plentiful in some countries that have never used it. It can be burnt, used to produce methanol, or gasified. In many aspects it is a traditional fossil fuel. Tar sands and oil shales are plentiful around the world, the biggest deposits in the Third World being in Venezuela. Like coal mining and oil production, these new energies need large centralised plants to compete with existing fuels.

So can renewable energy help the Third World? Not quickly and cheaply it seems. But appropriate application of some technologies – like solar power, biogas, mini-hydro and small windmills – would help if they were tailored to the needs of developing countries. And that calls for a village-by-village survey to discover what those needs are. It is unlikely that the nations at Nairobi will get down to such detail.

PART TWO

Plant power

The ability of green plants (and some bacteria) to "fix" carbon from the air and turn it into carbohydrates is the basis of our food chain. It is also the source of vast amounts of biomass material. As David Hall and Malcolm Slessor explain in the first article in this Part, the amount of material that photosynthesis produces every year could fuel the world 10 times over.

Of course it is not as simple as all that. Somehow, we have to turn this vast and dispersed resource into a usable fuel. We look at the technologies to do this, which involve combustion, or fermentation and distillation, in later parts of the book. But even ignoring the technical difficulties involved, exploiting plant power presents two problems. First, the resource is generally very thinly spread, so planners of biomass energy projects must take into account the amount of energy needed to harvest and transport fuel crops. Secondly, plants are not very efficient at converting energy: at best they convert about 2 per cent of the light that falls on the Earth's surface into chemical energy through photosynthesis.

There is no shortage of ideas for overcoming these problems. In "Cellulose from sunlight" Graham Chedd examines schemes such as vast plastic greenhouses, with CO_2-enriched atmospheres, and underwater seaweed farms. Another, which scientists working for NASA (of all people) have looked at, is to grow water hyacinths in sewage lagoons.

Ambitious high-technology schemes capture the imagination, but are unlikely to be economical sources of fuel crops in the near future. One way to tip the balance might be to grow certain fast-growing tropical plants, which use the so-called C4 mechanism of photosynthesis. Such plants include sugar cane, maze and papyrus. As Dr Peter Moore explains in "The varied ways plants tap the Sun", such mechanisms did not evolve by chance: they are an advantage in tropical conditions, but in the rest of the world normal "C3" photo-synthesis seems perfectly adequate. However there may be advant-ages in breeding C4 strategies into C3 plants.

Papyrus is an example of how productive a tropical plant can be. One hectare of the finest English pasture might yield 10 tonnes of grass a year. At Lake Naivasha in Kenya, the harvestable biomass from a hectare of papyrus reaches 32 tonnes per year. And it grows back in 9 months or so. The plant that fuelled the information technology of ancient Egypt is making a comeback as an energy crop.

The conclusion seems to be that photosynthesis can provide enough renewable material on which to fuel the world. But to make such a proposition practical, we must concentrate the potential, either in high-technology energy farms or by growing vast planta-tions of "high-speed" crops. Such a strategy raises environmental questions just as pertinent as those surrounding energy strategies based on coal or nuclear power. We shall turn to those in Part 3 when we look at the most dramatic manifestation of plant power – wood.

7

Self-sufficiency through biology

PROFESSOR DAVID HALL and
DR MALCOLM SLESSOR
25 July, 1976

Can plant scientists and microbiologists help to solve the food and energy crisis which developing countries now find themselves in? If suitable technology can be developed, how can it best be transferred to the developing countries? These questions were considered at a symposium in Sweden sponsored by the International Federation of Institutes of Advanced Study in conjunction with the UN Environment Programme.

The basic scenario discussed at the meeting centres on the use of solar energy to convert and store photosynthetic products as a basis for local self-sufficiency in food and fuel. Supplemented with carefully selected physical and mechanical systems, efficient integration of all these components on a local scale could provide the maximum sustainable level of per capita energy consumption. Decentralised societal and production organisations would be required to maximise this energy level.

Such down-to-earth plans, which involve persuading developing countries and local communities that they should try different approaches to solving problems to those advocated over the past 10 or 20 years, are not easy to translate into practice. However, plant and microbe-based systems exist that could be implemented at small and intermediate scales which would change energy consumption and food production in many countries. The basic science and technology already exists in various parts of the world, though often at a low level. What we now need are practical demonstration systems in these countries (and in the developed countries?) to show the energy and food gain achievable by going through the photosynthetic process and subsequently using the products directly or through microbial systems.

This means growing plants or algae solely for their food and/or fuel content, and also using the waste products of agriculture or urban environments through algalmicrobial systems. Pioneers in this field have been research workers in the United States, Israel, France and Germany, who have constructed systems in their own countries and elsewhere – for example, Thailand, Peru, South Africa, Mexico and India. There seems little doubt that with the correct input of research and development effort these systems could be viable in very many countries in the world and would be net energy and food producers. How to do so was one question on the agenda at the meeting.

The amount of energy available in photosynthetic products is considerable. In one year the amount of photosynthetically-fixed carbon in the world has an energy content of 3×10^{21} J – 10 times greater than the world's energy use in 1970. Of the photosynthetically-fixed carbon, only 0.5 per cent is used as food by the world's population of 4000 million people. The efficiency of photosynthetic conversion averaged over the whole planet is very poor – it represents a fixation of only 0.1 per cent of the total available light energy; good agriculture can give a 1 per cent efficiency. Thus the scope for improving the efficiency of photosynthesis and the utilisation of photosynthetic products seems to be enormous. There are problems, of course, but the plant is an ideal system for capturing and storing solar energy. Plants solved the energy crisis when the blue-green algae developed the process of oxygen-evolving photosynthesis about 3000 million years ago and it seems right that we re-examine how plants capture and store solar energy and that we implement this at our present level of scientific and technological sophistication.

The growth of algae on waste materials for food and fodder is a fairly well-advanced technology. Large-scale plants already operate in California, Israel and South Africa. The products can be methane gas, biomass for feeding to animals or for energy use, or protein for humans or animals – depending on the local economic and cultural situations and the requirements for waste disposal. These systems have been shown to be commercially viable and already seem to hold great promise in the sunnier parts of the world between 35° North and South.

Cellulose is probably the most abundant organic material on Earth, constituting up to half of the fixed carbon from photosynthetic processes produced every year, or about 10^{11} tonnes. Vast amounts are not used, however, because only ruminants can digest cellulose for conversion to food. Only recently have fungi and bacteria able to break down cellulose been considered for use in energy or food producing processes. This is now changing very quickly, especially since enzymes from fungi have been investigated which have been shown to break down cellulose to its constituent sugars. Once glucose is liberated, food, protein, organic substances, alcohol, fuels etc can be produced. Often the cellulose in cell walls of plants and algae is tightly bound through lignin components which hinder the breakdown of the cell walls, except when expensive milling processes are used. New processes, using enzymes which destroy the cross-linking of cell walls are being developed however, and could probably considerably change the economics of exploiting plant material for its cellulose.

Hence to the concept of *energy farming*, whereby plants are grown solely for their energy content; this seems to be more and more of a viable proposition in developing countries where land unproductive for raising food can be used to grow plants for their energy content: the plants are then degraded by various means to produce solid, liquid or gaseous fuel. This system is being investigated seriously in the US, Ireland, France, and Australia, but has immediate applicability in developing countries with favourable climates. Again, the economics and the net energy gain might vary from country to country, but by selecting appropriate plants, whether indigenous or imported, this could be a viable system. One of the advantages of energy farming is that the capital costs could be low and labour use could be high — a positive advantage in many developing countries.

The amount of nitrate and ammonia fertilisers used in agriculture throughout the world is enormous and it is now evident that the use of nitrogen fertilisers (and indirectly oil) has put many agricultural systems into the questionable area of not being net energy producers — or in fact just yielding only slightly more energy than is added in the agricultural systems. Of course, any net energy production in agriculture comes from the photosynthetic process. However, research on nitrogen fixation has changed dramatically in the past few years with the discovery of symbiosis in plants such as maize whose roots can contain nitrogen-fixing bacteria. Secondly, the nitrogen fixation genes have been shown to be a feasible way of incorporating nitrogen-fixing ability into ordinary plants. Another aspect of nitrogen fixation has been the realisation that the inherent limitation is often not the microbial system itself, but the lack of photosynthetic efficiency and the inability of plants to obtain sufficient amounts of carbon for the bacteria which fix nitrogen to grow efficiently. With these realisations, the practical application of science to the nitrogen fixation problem is really great.

More speculative implications of utilising plants or plant-based systems through more basic research were also discussed at the meeting. For example, plants contain chloroplast membranes which, in combination with a bacterial enzyme hydrogenase, can evolve hydrogen gas when illuminated. This system presently has severe stability problems but has three inherent capabilities which make it interesting and unique (analogous to the electrolysis of water). These are an unlimited input of energy (sunlight), and unlimited supply of substrate (water) to produce a non-polluting and storable source of energy (hydrogen). This system is now being actively investigated in a number of countries around the world; over the past three years the rates of hydrogen production attained have increased considerably.

Another idea mooted is the control of plant products by regulating their metabolism to produce more (or less) carbohydrate, fats, and proteins. As we know a lot about the metabolism of plants, there seems a possibility of regulating the plants to produce whatever end product we desire. Additionally, in bacterial photosynthesis there are a number of membrane systems which have very stable light-converting membranes which could be used to pump salts and produce electricity via a potential produced when light is shone on these stable membranes.

In the second half of the meeting, the advocacy of particular technologies gave way to thoughts on how they might be introduced to developing countries and on their economic and political repercussions. By this time it had become clear that to speak of developing countries as a single category was a gross simplification. Kuwait, for example, is planning a methanol/single-cell protein plant, while India needs low capital cost, low energy, intensive processes operable at village level. Such processes are by no means simple – indeed, their development calls for considerable scientific study of a high calibre – but they could be cheap and easy to operate.

There was some division of opinion as to what might be achieved by the implantation of processes into a simple village community. Were these methods simply to improve the use of land and waste while still using imported energy and fertilisers to drive their "system"? Or could such processes make it possible for communities to prosper, relying entirely on the scientifically contrived energy and food resources won from their land and creating a significant marketable surplus? The former was clearly possible but still left these communities at the mercy of external energy sources, presumably at rising prices. For this reason it was decided that a system simulation should be done using energy analysis techniques to test the potential for internal growth such as systems toward not only food and energy self-sufficiency but toward a growing economy.

8

The varied ways plants tap the Sun

DR PETER MOORE

12 February, 1981

Photosynthesis has apparent shortcomings which some tropical plants partly overcome in different ways. At first sight, these "super plants" would make excellent fast-growing crops for temperate countries. But nature, as usual, has arranged things well. In temperate climates, orthodox, slow, photosynthesis wins out.

The process by which green plants use the energy of the Sun to turn inorganic carbon dioxide gas in the atmosphere into organic molecules is crucial for all living things, except for a few specialist bacteria; it allows plants to grow, and ultimately provides food for animals. But this process, photosynthesis, is beset with difficulties, both logistic – the plant loses water as it takes in carbon – and (it seems) idiosyncratic; photosynthesis is depressed by oxygen, especially in hot conditions. To overcome these problems plants have evolved at least two strategies, which they superimpose on the basic mechanism of photosynthesis; yet both these strategies have disadvantages, the significance of which depends on the conditions. So which strategy a plant adopts (or whether it adopts one at all) depends not so much upon the kind of plant it is, as upon where it lives. Thus plants that are closely related may photosynthesise in quite different ways, while completely unrelated species – as different, say, as a palm tree and a creeping annual – may photosynthesise in the same way; and in general (although the correlation is far from simple) particular kinds of habitat will tend to favour one photosynthetic strategy above the others. To the botanist, the differences between the strategies are of course interesting; to the agriculturalist, they are of great economic significance.

The first – logistic – difficulty in carrying out photosynthesis is that of transpiration. If the carbon in carbon dioxide is to be trapped and turned into organic molecules ("fixed"), it must enter the plant through holes – so called stomata, situated mostly in the leaves; and if the gas is to enter the cells inside the plant, and hence reach the chloroplasts where photosynthesis takes place, the cells must be moist. Hence, as the plant takes in carbon dioxide gas, it tends to lose water by evaporation; and the more it loses, the more it must draw in from its roots. In dry conditions, unless the plant is specifically adapted, it will dry out and die.

The second problem in photosynthesis is photorespiration, a seemingly perverse process that was detected only about 20 years ago (the term photorespiration was first used and first described in detail by John P. Decker and Marco A. Tio in 1959). The term refers to oxygen competing with carbon dioxide and preventing its fixation; the net effect of photorespiration is that oxygen in the atmosphere depresses the rate of photosynthesis. Thus the natural atmosphere normally contains about 80 per cent nitrogen and 20 per cent oxygen (and only 300 parts per million of carbon dioxide). But if wheat is grown in the laboratory in an atmosphere containing only 2 per cent oxygen, then at a temperature of 25°C its rate of growth, which reflects the rate of photosynthesis, increases by about 20 per cent; in other words, under normal atmospheric conditions, wheat grows much more slowly than it could. It is important to specify the temperature because warm conditions stimulate photorespiration, and so the higher the temperature, the greater the depression of photosynthesis.

To understand photorespiration we should look at the basic process of carbon fixation – the method that all green plants possess, on which they may or may not superimpose other mechanisms. The simple overall formula of carbon fixation is $CO_2 + H_2O = (CH_2O) + O_2$. The group (CH_2O) is the basis of all sugars – the so-called pentoses, such as ribulose, which contain five such groups; the hexoses, like glucose, with six; and so on.

But in practice the splitting and recombination of carbon dioxide and water are highly complex. Carbon from CO_2 is first combined with a five-carbon compound, the phosphorylated sugar, ribulose bisphosphate (formerly called ribulose diphosphate), to form two molecules of a three-carbon compound, phosphoglyceric acid, or PGA. The three-carbon acid is then converted into a three-carbon sugar; and this three-carbon sugar (a triose) is then converted into other sugars (including ribulose, which can be used to capture more carbon) and hence into all other compounds (proteins, fats, and the rest) that the plant needs.

Photorespiration apparently takes place because oxygen interferes with the initial combination of CO_2 with ribulose bisphosphate. This reaction is orchestrated by the enzyme ribulose bisphosphate carboxylase. Oxygen apparently competes with CO_2 for sites on the enzyme, so that instead of combining with the carbon in CO_2 and then dividing into two three-carbon PGA molecules, the ribulose bishosphate splits to form one molecule of PGA and one of a two-carbon compound, phosphoglycollic acid (better abbreviated as glycollate). Glycollate is then broken down to make carbon dioxide again: and the whole process of glycollate formation and breakdown is stimulated by high temperatures.

Most plants fix carbon by the method of photosynthesis described above; and – because the first product of carbon fixation is the three-carbon PGA – are called C3 plants. But a C3 plant finds itself in a physiological Catch 22. In order to take in carbon dioxide it must sacrifice water by transpiration. In addition, its photosynthesis is stimulated by a rise in temperature but so too is photorespiration, and at high temperatures the carbon economy of the plant moves toward deficit; on still, hot days, the farmer's crop may actually become smaller. What we may call the "normal" C3 plant may find itself under particular stress in hot, dry conditions. The strategies that some plants adopt to overcome these stresses conform to two basic principles: one is to avoid opening their stomata during the day when stresses are at their greatest, the other is to reduce the

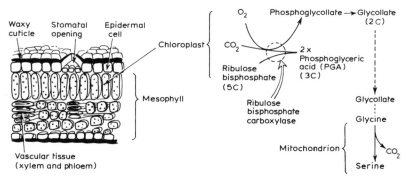

Figure 1. *Conventional photosynthesis – as performed by so-called C3 plants – is bedevilled both by water loss through the stomata (transpiration) and by photorespiration (as oxygen competes for ribulose bisphosphate). But C3 plants – like the oak – are highly successful.*

losses due to photorespiration. In both cases the carbon dioxide is at first fixed only temporarily, then released again and later fixed by the conventional route – ribulose bisphosphate and PGA.

Many species of plants combat dry conditions by the fleshiness of their leaves or, as in cacti, by their fleshy stems; and many of these so-called succulents reduce transpiration by keeping their stomata closed during the day and opening them only at night.

In these plants carbon dioxide, entering the leaves by night, first combines not with ribulose bisphosphate but with a three-carbon compound known as phosphoenol pyruvate (PEP), to form the four-carbon oxaloacetic acid (oxaloacetate) which is then converted to malate, also with four carbons per molecule. Malate can then be stored within the vacuoles of the leaf cells until daytime, when it is moved into the cytoplasm. There the malate is degraded to produce CO_2 once more – which is then refixed conventionally by combining with ribulose bisphosphate.

Such a mechanism clearly conserves water: the stomata are opened to receive carbon dioxide only when conditions are least likely to cause evaporation. This system was first described in 1947 by Meirion Thomas, in the family Crassulaceae, which is represented in Britain by the stonecrops; hence the system is known as

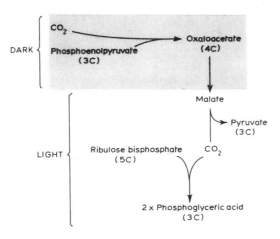

Figure 2. *CAM plants, such as cacti, reduce transpiration losses by opening their stomata to take in carbon dioxide only at night. They fix the carbon temporarily and then, by day (and with the stomata closed), re-fix it permanently by the conventional route. CAM plants do well in the desert, but they grow slowly.*

C

Crassulacean acid metabolism, or CAM. But we now know that CAM occurs in 21 families, representing both main groups of flowering plant: the monocots (the group that includes grasses, aloes and orchids) and the dicots (which include all other flowering plants from cacti and asters, to stonecrops). Many, perhaps most, of the succulents that practise CAM are also able to fix carbon by day by the normal route, when conditions are moist.

In 1965 Hugo Kortschak and his colleagues in Hawaii showed that in sugar cane the initial product of carbon fixation was not PGA; and the following year M. D. Hatch and C. R. Slack in Australia worked out the full mechanism. Carbon dioxide first reacted with the three-carbon PEP to form oxaloacetate as in CAM plants; but in cane (unlike the CAM plants) this process was not restricted to the night. In CAM plants the two processes – the initial fixation to PEP, and conventional fixation – are separated by time. In cane there is no delay between the two processes; but they take place in different parts of the plant. Hence, the initial fixation to oxaloacetate takes place in the cells in the middle layers of the leaf (the mesophyll) and the oxaloacetate so formed is then converted to four-carbon malate or aspartate. But the four-carbon products are then transported to specialised cells which surround the vascular

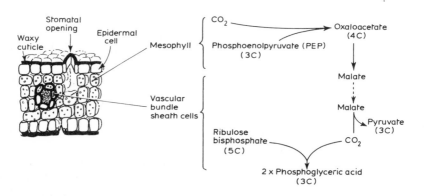

Figure 3. *C4 plants reduce losses due to photorespiration by fixing carbon temporarily, then releasing it, then re-fixing. This resembles the strategy of CAM plants, but is all done by day; special cells around the vascular bundles refix the carbon after temporary storage.*

bundles in the leaves, the tubes and columns of cells that convey water and nutrients through the plant. These specialised cells release CO_2 from the four-carbon molecules, and re-fix it with ribulose bisphosphate.

In plants that possess this so-called C4 mechanism, the specialised vascular bundle sheath cells in the leaves have a distinctive structure and contain conspicuous chloroplasts; indeed, C4 plants can often be recognised by their anatomy, without detailed biochemical analysis.

Overall, although photorespiration must take place in C4 plants, its effects are not observable. Neither high leaf temperature nor oxygen depresses these plants' assimilation of carbon. The reason, probably, is that the site of conventional carbon fixation is well insulated in C4 species: any CO_2 formed within the vascular bundle cells as glycollate is formed and degraded is likely to be trapped by another PEP molecule on its way through the cytoplasm. In addition (presumably in part because they avoid the losses due to photo-respiration) C4 plants can maintain a positive carbon balance even when the concentration of CO_2 is very low – less than 10 ppm, which is about one-thirtieth of the normal atmospheric concen-tration. Finally, C4 plants can utilise very intense light.

With such advantages it is hardly surprising that under ideal conditions C4 plants can grow very rapidly. Sugar cane and maize, which are grasses, and papyrus (*Cyperus papyrus*) which is a sedge are good examples. *Pennisetum*, a tropical C4 grass, can yield 85 000 kg/ha of forage in a year; compare this with the 30 000 kg/ha of rye grass (*Lolium perenne*), the most common forage grass in Britain, which uses the C3 mechanism. In truth, this difference cannot be ascribed simply to the difference between C3 and C4; tropical grasses have a longer growing season than that of temperate species. Even so, the C4 mechanism, under appropriate conditions, has clear advantages; and although it was first described in sugar cane (a grass) it has now been observed in 19 families of flowering plant, as well as in a fern and a gymnosperm.

Most C4 species live in hot, dry conditions. The British Isles have only a few C4 species, including the salt-marsh grass *Spartina anglica* and some maritime dicots such as *Sueda fruticosa* (shrubby seablite) and *Salsola kali* (prickly saltwort). In California more than 4 per cent of the dicots and 80 per cent of the grasses have adopted the C4 strategy. In the US the proportion of C4 species among the grasses diminishes toward the north: the closer the plants are to the Equator, it seems, the greater the advantage of the C4 mechanism.

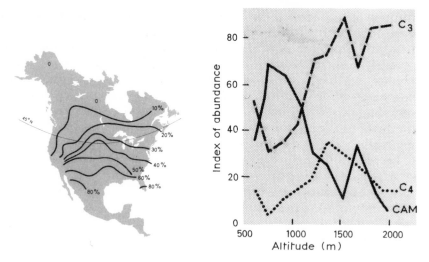

Figure 4. *C4 plants do best in the tropics: the percentage of grasses using the C4 mechanism in North America diminishes toward the north; and in a study in Texas, CAM plants give way to C4 species, and the C4s lost out to C3 types as altitude increased.*

In addition, C4 plants often seem to grow most quickly when C3 plants are growing slowly and vice-versa. Thus in the southern Great Plains region of central North America, 11 of the 15 species of C3 grasses attain their maximum growth during the cool season, between September and June. All of the 34 C4 species grow best in the hot season, between April and October. In the aridity of Death Valley, California, almost all of the species active in summer are C4 plants, though there are some interesting C3 exceptions.

Research in Texas has shown that the distribution of C3 and C4 and CAM plants is related to altitude. CAM plants dominate the desert communities at low altitudes; C3 species are at a maximum at intermediate heights; and C3 species dominate the high-altitude communities. As altitude increases in this area, so conditions become cooler and less arid. In Kenya, too, there are no C3 grasses below 2000 m, and no C4 grasses above 3000 m.

Photosynthetic strategy is by no means the only factor that determines where a plant lives, but it is important. All three mechanisms, C3, C4 and CAM have both advantages and disadvantages, and a plant can compete succesfully only when the benefits of its own photosynthetic preference outweigh the snags. CAM plants seem best adapted to very arid conditions. Their method of conserving water by closing their stomata by day is valuable in the desert, but it severely reduces their ability to take in and assimilate carbon. Hence, CAM species grow slowly and, in less extreme conditions, compete poorly with C4s and C3s. The CAM plants come into their own only when conditions are so dry that the more straightforward methods of fixing carbon lead to excess loss of water. C4 species generally tolerate dry conditions and high temperature better than C3 plants do. A high leaf temperature does not depress their photosynthesis, and whereas C3 species transpire about 500–700 grams of water for every gram of dry matter they produce, C4 plants lose only 250 to 400 grams of water per gram of dry matter. So in hot, fairly dry conditions – and in saline habitats such as the British salt-marshes, where there are also problems of water balance – C4 plants can grow quickly and present formidable competition to the C3 species.

But at temperatures below 25°C, C4 plants may lose out to C3 species. They are more sensitive to cold, perhaps because some of their enzymes function less well at low temperatures than those of C3 species. In addition, the C4 system needs more energy to run – energy expended in regenerating PEP. (Specifically, the C4 system requires two to three extra molecules of adenosine triphosphate (ATP) for every atom of carbon fixed.) Thus to some extent C4 plants are victims of their own system. They can grow extremely quickly; but *unless* they grow extremely quickly they are not at an advantage.

Where summers are dry and hot, the agriculturalist can tap the advantages of C4 plants. Many of the tropical and sub-tropical species (especially grasses) that have been brought into cultivation are C4 plants, including maize, millet, sorghum and sugar cane. Most temperate crops, such as wheat, barley, beans, peas and potatoes, are C3. Perhaps we should introduce more C4 species into those temperate countries that are sometimes short of water in summer. Maize has indeed been introduced into southern England – grown mainly for forage (cattle eat the entire plant, leaves and all) rather than for grain; and some plant physiologists, such as Harold Woodhouse at the University of East Anglia, have stressed the possible advantages of introducing C4 grasses into England's southern pastures.

On the other hand, some C4 plants have become serious agricultural weeds in warm climates, including cockspur grass

(*Echinochloa crus-galli*), bermuda grass (*Cynodon dactylon*), which is also used for lawns in many areas, and purslane (*Portulaca oleracea*). All those plants originated in the tropics, but are now widespread.

In practice, whether agriculturalists will be able to make significant strides by introducing more C4 species, or, as some have suggested, by trying to breed the C4 strategy into C3 types is a moot point. Although the C4 strategy has obvious theoretical advantages, the C3 species compete remarkably well under most conditions; and, in general, the limits to growth and profitability of crops usually have far more to do with the supply of water and of fertiliser (the right amount at the right time) and with the quality of husbandry, than with the plant's own photosynthetic strategy.

We are left, however, with a botanical conundrum. Why, having produced the miracle of photosynthesis, did nature then sabotage the system by allowing it to be depressed by oxygen? In fact, photorespiration may be an accident. The mechanism of photosynthesis almost certainly evolved in a primitive atmosphere in which oxygen was scarce: atmospheric oxygen is largely a by-product of photosynthesis. Current oxygen levels were probably achieved by Carboniferous times (350 million years ago) and from then on the photosynthetic green plant would have been embarrassed and rendered less efficient by the by-product of the very system that had made them so successful.

9

Cellulose from sunlight

GRAHAM CHEDD
6 March, 1975

How can scientists improve the amounts of biomass that plants produce? Ideas abound – from giant greenhouses with atmospheres enriched with carbon dioxide to underwater kelp farms. But no-one has yet demonstrated that such projects would be an economical way of producing energy crops, and of course carbon dioxide is only one input among many. Most plants, most of the time, are more likely to be short of nitrogen or water.

Agriculture is the only industry that today depends upon incident solar radiation. But even agriculture, at least as it is practised in the West, relies on a heavy investment of fossil fuel to harvest sunlight efficiently. Indeed, it is the generally low efficiency of photosynthesis in turning solar energy into chemical energy that is the biggest problem behind the otherwise appealing idea of utilising plants as a major source of energy. At this year's meeting of the American Association for the Advancement of Science James Bassham of the Laboratory of Chemical Biodynamics, Lawrence Berkeley Laboratory, pointed out that photosynthesis produces some 155 billion tonnes dry weight of organic matter, largely cellulose, each year; but that the efficiency of conversion of total solar irradiance into cellulose is commonly around a few tenths of 1 per cent. This means that large acreages would have to be planted as an energy crop – which immediately raises the question of where the land is to come from at a time when every plantable square metre of the world should be given over to raising food crops.

One possible way out of the impasse is to find a plant with a much higher conversion efficiency. Bassham discussed the so-called C4 plants such as sugar cane and sorghum, which under optimal conditions have reached conversion efficiencies approaching 3 per cent.

Ordinary plants fix carbon via the Calvin cycle, by which carbon dioxide is reduced to sugars. The rate of this reaction is limited by the supply of carbon dioxide to the plant's chloroplasts; even when plenty of sunlight, water and nutrients are available, the amount of carbon fixed is limited to the rate of supply of carbon dioxide. C4 plants possess a chemical pump, which keeps the chloroplasts primed with carbon dioxide, and so allows a more effective utilisation of the high sunlight intensities of the tropics. But Bassham pointed out that even non-C4 plants can reach higher conversion efficiencies if the atmosphere in which they grow is enriched with carbon dioxide, to a level of say 0.2 per cent from the normal 0.03 per cent. (This technique has already been used for some years for high-value greenhouse produce.)

Bassham envisaged large closed-cycle plastic greenhouses in the deserts of the US South-West as one possible solution to economic cellulose farming. These greenhouses, using recycled water and a CO_2-enriched atmosphere, might house a crop such as alfalfa. US Department of Agriculture scientists have already developed methods for extracting protein from alfalfa leaves in a form suitable for human nutrition. The present process does not extract all the protein, and the residue is used as a cattle feed. Bassham foresaw technological developments that might allow the extraction of virtually all the protein from alfalfa, leaving a cellulosic fibre. Suppose, he argued, that with the aid of a CO_2-enriched atmosphere such a crop could be boosted to yield 80 tonnes dry weight per hectare per year. Suppose too that from this 10 tonnes of vegetable protein and 25 tonnes of cellulose and sugars could be recovered. At $2 a kilogram the protein would be worth $20 000 – an order of magnitude less than that considered economic for greenhouse crop production at the moment, but that might change. The amount of sunlight fixed as cellulose and sugars would be around 1.5 per cent, a respectable figure. Bassham can see the flaws in his proposal (a little energy input–output analysis might, one suspects, destroy it altogether) but it has the virtues of producing cellulose alongside a valuable food product – without competing for present agricultural lands.

A more straightforward if brute-force solution to the problems of energy farming was proposed by Clinton Kemp of Inter-Technology Corporation. He listed the requirements of a source of major energy. It should be: capable of storing energy for use at will; high in thermodynamic availability; forever renewable; ecologically inoffensive; widely available in the United States; dependent on existing technology; operable on a large scale in a decade or two; "tolerably" priced; and free of horrendous hazard potential. His solution: energy plantations planted with "Btu bushes".

Kemp has done rough energy input–output and cost analyses for his proposal, and came up with some intriguing figures. For instance, if maize were raised for energy instead of animal feed the costs would be comparable to those presently paid for fossil fuels in the US. If forest conifers were used, planted at some 12 000 trees per hectare and harvested about once every 12 years the costs would be roughly doubled, because of slower growth rates and very large inventories, to around $3.00 per million Btu (1 Btu = 1055 Joules) of useful fuel. But perennials have one invaluable advantage over annuals: they can be harvested throughout the year in response to demand, avoiding the huge storage problems of an energy crop that could be harvested only with the seasons. So Kemp and Inter-Technology Corporation have turned in their search for suitable Btu bushes to fast-growing deciduous trees which re-sprout from the stump when cut, avoiding the need for re-planting every year.

In experimental plots on abandoned farmland in central Pennsylvania they have been growing hydrid poplars from a number of promising clonal lines. One hybrid (Clone 388) planted at some 7500 trees per hectare and producing about six or eight harvests per planting, has costs in the range $1.25 to $11.45 per million Btu (as compared with oil's present cost of $1.97 per million Btu and coal's $1.31 per mBtu). Such a plantation could produce around 240 mBtu per hectare per year, at a conversion efficiency of 0.6 per cent. It would grow on marginal land. Some 1200 hectares would be needed to fuel a medium-sized 400 MWe power plant; "as a matter of fact", according to Kemp, "to supply the fuel from energy plantations for all the generating capacity presently installed (in the US) would require less than 160 million acres [64 million hectares] even at the rate of conversion of solar radiation to fuel value as low as 0.4 per cent". This average is less than a third of what Inter-Technology estimates is available for Energy Plantations. "We are only at the foot of the learning curve", says Kemp. "If we could get to a 1 per cent efficiency, we could be exporting the stuff."

One method Kemp considers for using the energy stored in his Btu bushes is to convert it to methane via anaerobic digestion. The development of a "methane economy" based on the anaerobic digestion of plant matter has been urged by Donald Klass, assistant research director of the Institute of Gas Technology; and while Klass was not a speaker at the symposium, his ideas are clearly gaining support. For example, William Oswald of the Civil Engineering Department at the University of California, Berkeley, discussed his experiments on the digestion of algae grown on sewage stabilisation ponds. His project began by accident, when nutrient-rich stabilisation ponds built to protect the Napa river in California developed dense algal blooms. Oswald is testing several easily harvested varieties of algae to see if they will grow on fresh sewage and digest efficiently to produce methane.

A project somewhere between Oswald's and Kemp's in concept but rivalling Kemp's for audacity is being tested by, of all people, the US Navy, at the Naval Undersea Center, San Diego. This is the Ocean Food and Energy Farm, discussed at the symposium by Howard Wilcox

Wilcox plans to solve the land-shortage problem by going to sea, there to grow on underwater rafts macroalgae-seaweed supplied with nutrients not by sewage but by deep, cold, nutrient-rich water brought to the surface by artificially induced upwellings. Seaweed contains little cellulose, but could be processed by chemical or bacterial action to produce methane plus other products such as fertiliser, ethanol, lubricants, waxes, plastics, fibres – indeed, a complete spectrum of petrochemical-type products. Furthermore, the seaweed could also be used, directly or indirectly, as a food source.

The navy has already installed a 3-hectare experimental underwater farm of kelp, one of the world's fastest growing plants, in about 100 metres of water just off San Clemente Island. California already has a well-developed kelp industry based on natural kelp buds, and on this experience and their own from the experimental farm Wilcox and his colleagues have begun food production and fuel conversion experiments. From these small beginnings, Wilcox can envisage a dramatic future:

"Assuming a 2 per cent conversion efficiency for converting solar radiation into the stored energy of seaweed compounds, 5 per cent conversion efficiency for the production of human food from the seaweed, and 50 per cent conversion efficiency for the production of other products from the seaweed, the marine farm is conservatively projected to yield enough food to feed 3000 to 5000 people per square mile of ocean area which is cultivated, and at the same time to yield enough energy and other products to support more than 300 at today's US per capita consumption levels, or more than 1000 to 2000 people at today's *world average* per capita consumption levels. Since the oceans conservatively estimated, appear to contain some 300 million square kilometres of arable surface water, this means that marine farms could conceivably support a human population ranging from 20 to more than 200 billion people, depending on the degree of affluence assumed."

A rather more immediately practical source of methane is urban solid waste. D. I. Wise of Dynatech, a Cambridge, Massachusetts, research and development company, presented results of an engineering analysis of a plant processing 1000 tonnes of waste a day and serving a metropolitan population of about 500 000 people. Gas would be produced by anaerobic digestion from this approximately 280 tonnes of organic material that 1000 tonnes of waste would contain. The computer analysis finishes up with a price for the gas of almost $55 per million cubic metres – more than twice the current cost. But if the analysis includes credits for the waste disposal, the usable sewage sludge and the scrap iron the process would recover, the cost of the gas drops sharply to a little over $18. The final price takes into account the plant's own energy consumption, of roughly 37 per cent of the gas produced. As Wise pointed out, "such a system may be a 100 per cent solution to the solid waste problem, but it would contribute only about 5 per cent to the annual US production of natural gas". On the other hand, 5 per cent is 5 per cent; and in states such as Massachusetts, on the end of the natural gas pipelines, the percentage of annual gas consumption capable of being generated from solid waste reaches about 28 per cent. Not surprisingly, Dynatech's analysis is receiving a great deal of attention from municipalities already struggling hard to solve their solid waste disposal problems.

Perhaps the most striking presentation in the symposium, both for its promise and its practicability, was also superficially the most bizarre. The key to the research of Leo Spano and his colleagues at the US Army Natick Laboratories in Massachusetts is a fungus first isolated from a rotting cotton cartridge belt in the jungles of New Guinea in the closing weeks of the Second World War. The army's Natick labs have responsibility for all the consumables used by the US Army and airforce, and one of the problems they have studied over the years is the protection of clothing from rot. In the past 30 years, the labs have collected some 13 000 microorganisms that can live on cellulose by producing cellulases, enzymes capable of breaking the cellulose polymer back into the glucose from where the original plant constructed it. These enzymes are usually secreted by the microorganisms into their environment.

When, in 1971, the Natick labs were given the task of doing

something about the mounting waste disposal problems on US Army bases, someone remembered their star cellulase-producing microorganism, the fungus from the cotton belt, *Trichoderma viride*. *Trichoderma*'s special trick is to produce a cellulase rich in a component capable of breaking down crystalline and generally insoluble cellulose. (Most commercially produced cellulases are obtained from *Aspergillus niger*, which has only a trace of the so-called C_1 enzyme.) Spano and his colleagues derived from the original *Trichoderma* strain mutants that produce two to four times as much cellulase as the wild type, and believe they have not yet reached the upper limit.

The process developed at Natick is simplicity itself. The *Trichoderma* is grown in a culture medium containing spruce pulp and nutrient salts. The culture is then filtered and the solids discarded, leaving a clear straw-coloured enzyme solution with a marked resemblance to American beer. The cellulose solution and milled newspapers are then placed together in a reaction vessel at atmospheric pressure and 50°C. The product is crude glucose syrup; the unreacted cellulose and enzyme are recycled. The yield of glucose is about 50 per cent of the original cellulose. It can be used in chemical, or microbial fermentation, processes to produce chemical feedstocks, single-cell proteins, solvents or fuels such as ethanol.

Spano and his colleagues have tested their process for a variety of cellulose sources – mostly waste-paper – pretreated in a variety of ways. Among the most suitable, Spano was delighted to recount, were Pentagon documents, which went through the process in a way "you just wouldn't believe", producing pure sugar in just a few hours of digestion. About to come into operation at Natick is a full-scale pilot plant with a capacity of up to 500 kg of cellulose per month and capable of being stretched to a through-put of some 2000 kg. The army process is receiving a great deal of attention, both nationally and internationally. Several large chemical companies are seriously considering the production of ethanol by the fermentation of glucose – the way it used to be done, before ethanol was made from oil – and are acknowledging Natick's lead in the production of glucose from cellulose.

Almost all proposals for making fuels with photosynthesis – be the eventual product methane or ethanol – start with the reduced carbon products made in the plant. Melvin Calvin, after whom the chemical cycle in which the carbon is reduced is named, has for several years championed an alternative: intervene in the energy capture step of photosynthesis and tap off the energy directly, either as electricity or hydrogen. Calvin and his group are working on these ideas at Berkeley.

Another team interested in the production of hydrogen from the biophotolysis of water is led by Donald Krampitz at Case Western Reserve University, Cleveland, Ohio. Krampitz told the symposium of his group's work to find a chemical that will "store" the electrons produced in the first stages of photosynthesis so that they can then be used to reduce protons to free hydrogen. Krampitz has been experimenting with a dye, methyl viologen, in a system combining the photosynthetic apparatus from a blue-green alga and a hydrogenase preparation (to catalyse the reduction of protons to hydrogen) from *Escherichia coli*. Light shone on the system produces hydrogen, at least proving that the biophotolysis of water is possible. Krampitz believes that "a goal of 10 per cent conversion of photosynthetic energy into hydrogen is attainable".

One of the main arguments against the widescale use of solar energy has always been that the Sun does not always shine when you need it; that its energy has somehow to be stored. A corollary has been that what we are running out of is clean liquid and gaseous fuels rather than energy *per se*, and that you will never be able to power a car with solar energy. Using photosynthesis to turn sunlight into cellulose, and subsequently other enzymatic reactions to turn the cellulose into methane or alcohol, solves both these problems while not bringing in its wake any obvious environmental penalties. It could also relieve the strain on hydrocarbons as a source of petrochemicals. And the appropriation of photosynthesis to produce hydrogen directly at efficiencies approaching 10 per cent could have truly startling consequences for the world's energy supply.

Papyrus plants near Lake Naivasha, Kenya.

10

Papyrus: a new fuel for the Third World

MICHAEL JONES
11 August, 1983

The most unlikely plants turn out to be good biomass fuels. One such is papyrus, one of the most productive plants in the world. Exploiting papyrus properly may help many African countries overcome their shortages of fuel.

Papyrus is the plant that fuelled the information technology of ancient Egypt; the Egyptians made from it the writing material that supported a full-blown civil service and all the trappings of civilisation. This ancient crop now has a new future. Technology originally designed to convert peat into an efficient solid fuel in Ireland may soon be turning vast African swamps of papyrus into vital fuel for hard-pressed Third World countries.

Papyrus (*Cyperus papyrus*) provided the principal writing material in much of the ancient world for more than 5000 years, and it was cultivated by the Egyptians on the banks of the Nile. But the growing use of parchment and vellum, both made from animal skins, during the first eight centuries AD gradually ousted papyrus from its supremacy. The pressure to use the fertile swamp soils for other crops was then so intense that the papyrus was cleared and eventually completely disappeared from Egypt. In recent years, however, papyrus has been reintroduced to a limited extent; mainly to satisfy tourists who wish to see the ancient art of paper-making.

Papyrus is the largest of the sedges, a group of plants closely related to the grasses. The stems or "culms" grow up to 5 metres and are topped by a large feathery crown called the "umbel", which is the flower-bearing structure or inflorescence of the plant. However, flowers are either absent or largely inconspicuous and the many thin green rays that form the bulk of the umbel are the principal photosynthetic organs of the plant.

Most papyrus is now concentrated in east and central Africa. One million hectares of the Sudd swamps are covered with papyrus. It forms a virtual monoculture, excluding all potential competitors apart from an occasional climbing hibiscus.

Papyrus grows in two types of location: in the Sudd the floating mats occur at the edge of many shallow lakes, or even fill them completely. In the second type, primarily in southern Uganda and Rwanda, the swamps cover the floors of valleys which carry the rivers that for the most part feed Lake Victoria. In both settings the papyrus lies on deposits of peat built up over many centuries from decomposing papyrus and other vegetation.

We have learned a lot more about the ecology and physiology of the papyrus recently, and one of the most striking discoveries is that it is among the most productive plants in the world. The harvestable standing biomass reaches a maximum of 32 tonnes per hectare (t per ha) at Lake Naivasha in Kenya, although harvestable biomass varies from place to place. (Biomass is the total mass of growing material including both living cells and dead components, such as wood.) By comparison, grass grown on the finest English pasture yields around 10 t per ha. In general, the swamps at higher altitudes, near the upper limit of papyrus of about 2000 metres, have the highest biomass; the higher altitude swamps, such as those on Lake Naivasha may be able to maintain larger amounts of standing vegetation because the lower temperatures at high altitude reduce the losses of carbon through respiration. As temperature falls, the rate of respiration declines more rapidly than the rate of photosynthesis so the plant ends up with a net gain of carbon. Papyrus will regrow rapidly after it has been harvested and can probably regain its original biomass within 9 months to a year in most cases. This would mean that the annual production of the most productive swamps is about 30 t per ha, a value that compares very favourably with forests as a source of biomass.

Cut papyrus is able to regrow so rapidly because the plant has a plentiful supply of water, and because it belongs to a group of plants which have the C4 pathway of photosynthesis. Several other crop plants, such as maize and sorghum, have this type of photosynthesis and they are all particularly productive under tropical conditions. The roots of papyrus also harbour nitrogen-fixing organisms which provide nitrogen in a form that the plants can take up as nutrient and is equivalent to the amount added to moderately fertilised grassland in Europe.

Our increased knowledge of papyrus has come at a time when we have become acutely aware of the "other energy crisis", particularly in central Asia and Africa. This crisis is due to the rapid disappearance of wood as the major source of fuel, either burned as wood or in the form of charcoal, and the lack of alternative sources of energy. Much effort is being directed toward finding new sources; for example an American and Irish aid programme is developing the excavation of peat in the central African state of Burundi (which has reserves of 500 million tonnes). But there are problems in draining the bogs to extract the peat; so papyrus, which in many cases grows on top of the peat, is an obvious alternative. The real advantage of papyrus is that it is truly renewable, whereas exploiting peat leaves large areas of open water suitable for little apart from fish farming. The feasibility of using papyrus as a source of fuel is now being investigated with funds provided by the Irish government's bilateral aid programme.

Papyrus is clearly productive enough to be a source of fuel, but when it is simply air-dried it is not dense enough to provide the concentrated heat required for cooking, for instance. The breakthrough has come with the idea of compressing the air-dried papyrus to about one-twentieth of its original volume into a briquette, using a machine similar to those used in developed countries to produce briquettes from waste straw and wood chippings.

James Martin, a retired engineer from the Irish Peat Board, has been the driving force behind this development over the last two years. He has demonstrated that chopped papyrus that has been air-dried can provide a combustible briquette with a density slightly greater than that of wood. The resulting fuel produces little smoke on burning and it has a low ash content – an admirable substitute for the rapidly disappearing charcoal which is the main source of energy in many African cities.

Initial attempts to produce briquettes from papyrus are being carried out in Rwanda; the briquettes will supply its capital, Kigali. Large areas of swamp (more than 5000 ha) lie within 40 km of Kigali, which reduces the cost of transport, one of the main determinants of cost of fuels. The papyrus will be cut by hand as the swamps are accessible for at least 7 months of the year, when the water levels are low. The papyrus will then be dried and compressed into briquettes at the edge of the swamp.

Large-scale exploitation of these swamps clearly has some dangers; not least that posed by removing large amounts of biomass from an ecosystem with a tight cycle of nutrients similar to that of tropical rainforests. At first, however, the briquetting factory will need only relatively small areas; about 60 ha should supply the equivalent of one-fifth of Kigali's requirements for charcoal.

We still have to determine just how much can be cropped year to year from these swamps and the effect that continued harvesting will have on their ecology. For instance, production might fall after some years because of lack of nutrients. And opening up the canopy by removing papyrus will allow more light to penetrate and might increase the proportion of other plant species in the swamp.

Papyrus swamps are not easy to investigate. They are often difficult to penetrate and harbour the vectors of diseases such as schistosomiasis (bilharzia) and malaria. But many aspects of the ecology of papyrus swamps have been studied, as very ably summarised in *The Inland Waters of Tropical Africa* by L. C. Beadle, formerly professor of zoology at Makerere University, Uganda. One main feature is the almost complete dominance of the papyrus plant itself, which usually accounts for more than 95 per cent of the total plant biomass. The papyrus vegetation is so dense that only specially adapted species such as climbers (particularly *Hibiscus diversifolia*, *Ipomea cairica* and *Melanthera scandens*) can survive, using the papyrus stems to find more light near the top of the canopy. Ferns, adapted to low levels of light, can also live in papyrus swamps.

Another feature of papyrus swamp is that it is almost totally depleted of oxygen (that is, is anoxic) and is reducing. This results from the rapid growth, death and eventual decomposition that occurs continuously in the water and mud around the papyrus mat. Only anaerobic microorganisms which survive in the absence of oxygen can thrive – for most organisms this is a hostile environment – although Beadle showed that the range of fauna is richer than might be imagined. Some animals have accessory respiratory organs for breathing air, as in the lung-fish *Protopterus* and *Polypterus*: but most of the animals living in the swamp water and mud do not breathe air continuously and they are therefore likely to have developed a metabolism that is at least partially anaerobic.

The roots of the papyrus are of course also in an anoxic environment but the plant has overcome this by developing a system of large interconnecting spaces between the cells. Oxygen diffuses in through the pores (stomata) in the aerial parts of the plant and moves down to the roots where it is consumed in respiration.

Papyrus swamps also provide a habitat for terrestrial animals, including insects, birds and a few mammals; the largest of these is the hippopotamus which makes very prominent tracks through the swamp. Swamps are rarely the preferred habitat for any species although they do form a very important wildlife refuge, particularly in the more densely populated parts of Africa such as Rwanda.

But we will not know what effects the regular removal of papyrus from the swamp will have on the ecology until we have done more work on managed swamps. One thing is clear; the complete removal of papyrus swamps could be ecologically disastrous. There are many reasons for this but probably the most important is that the total mass of animals in the swamp is small compared to the plant biomass. As a result, much of the decaying plant material leaves the swamp to become organic nutrients for plants and animals elsewhere. The amount of this detritus that is exported depends largely on the rate at which water moves through the swamp; but the swamp also acts to control this rate of movement, being in effect a natural area of storage and discharge. The productivity of neighbouring ecosystems are therefore likely to depend to some extent on the presence of papyrus and its removal could clearly affect their productivity. Also, the swamps act as a giant silt trap for waters flowing through them and therefore reduce the amount of silting downstream. And the effect of by-passing a large area of the Sudd with the Jonglei canal, now under construction, is still far from clear.

Regular and controlled cropping of the papyrus should not harm the ecology of the swamp and could even be beneficial. Preliminary work has shown that letting in more light by harvesting papyrus allows new species to invade, which could reduce yield of papyrus but may also produce a better environment for wildlife. And we already know, from the recent work of Dr John Gaudet in Uganda and Kenya, that these swamps can remove large amounts of nutrients from the water that flows through them and so may help reduce the danger of eutrophication – the phenomenon by which large bodies of water become too rich in nutrients, and fill up with unicellular algae which obliterate other life forms. The swamps could even "clean up" the liquid waste produced by towns: perhaps the extra need that cropped papyrus swamps have for nutrients would be an advantage – a bonus favouring harvesting for fuel.

Clearly, the prospects for a new future for papyrus seem good in a part of the world which has few natural resources, and where there is a very real danger of energy starvation.

11

Don't waste waterweeds

BILL WOLVERTON and
REBECCA C. McDONALD
22 February, 1976

By 1976, biomass was a respectable subject for study and scientists were looking for new plants to exploit. The water hyacinth, which clogs rivers throughout Africa and Asia as well as in the southern US, could be a source of biogas, fertiliser and food.

Japanese exhibitors coming to the 1884 Cotton Exposition in New Orleans brought with them an attractive water hyacinth they had collected from a river in Venezuela. It escaped – and today many of the world's major rivers between 32°N and 32°S latitude are infested with this beautiful pest (*Eichhornia crasspipes*).

Highly prolific and reproducing mainly by vegetative offspring, water hyacinths can double in number every 8–10 days in warm nutrient-enriched waters, forming huge solid floating mats. The plant consists of a fleshy, verticle stem called a rhizome, from which roots, leaves and flowers develop. The rhizome floats just beneath the water, where it is protected by shields of folded leaves, remaining alive when frost kills the surface leaves. Unless the water freezes, viable rhizomes will respond to warm water (10°C or greater) by producing new growth.

After its introduction into the United States, water hyacinths reproduced so rapidly that within only a few years rivers and streams in Louisiana and Florida were completely clogged with mats of plant material. The US Army Corps of Engineers was given the task of controlling this new noxious weed. In the early 1900s it started using sodium arsenite, but the hazards associated with this chemical were eventually recognised and in 1937 it was abandoned in favour of mechanical harvesting. After 10 years of a losing battle, the Corps of Engineers turned to the plant hormone 2,4-D when it became available in the late 1940s. For over 25 years this herbicide alone has been used in an effort to control *Eichhornia*.

In 1973 water hyacinths infested about 200 000 hectares of Louisiana water, and by 1975 the affected area had grown to more than 400 000 hectares. The Corps of Engineers and the US Department of Agriculture are evaluating biological control in the form of microorganisms and insects, combined with mechanical and chemical methods. But until recently no serious attempt was made to exploit fully the desirable qualities of *Eichhornia*.

We have been so busy trying to destroy this beautiful plant that we failed to recognise fully its potential value in pollution control, and as a source of energy, food and livestock feed, fertilisers and other products. Cursed for so many years, the water hyacinth is now beginning to gain respectability by offering relatively simple and economically attractive solutions to some of mankind's most pressing problems.

The US National Aeronautics and Space Administration (NASA) has demonstrated the remarkable mineral and nutrient uptake of water hyacinths at its National Space Technology Laboratory (NSTL). Grown in warm, enriched domestic sewage *Eichhornia* produces over 17.8 tonnes of wet biomass per hectare per day. Such plants contain 17–22 per cent crude protein, 15–18 per cent fibre and 16–20 per cent ash. Growth rate studies suggest annual production rates of 712 tonnes of dried plant material per hectare.

Raw sewage from small communities in South Mississippi contains an average of 35 milligrams per litre of nitrogen and 10 mg/l of phosphorus. On this basis, a half-hectare lagoon covered with water hyacinths, with a minimum sewage retention time of 2 weeks, should be able to purify, to acceptable levels, the daily wastes of 1000 people. An experimental water hyacinth lagoon did indeed reduce pollutant levels by 75–80 per cent.

Water hyacinths could also prove useful in treating effluents polluted with toxic heavy metals. In static laboratory experiments, *Eichhornia* rapidly absorbed gold, silver, cobalt, strontium, cadmium, nickel, lead and mercury. Ninety-seven per cent of the cadmium and nickel was concentrated in the roots within 24 hours – although the roots constitute only 18 per cent of the total dry weight. Water hyacinths can also absorb or metabolise phenols and other trace organic compounds of the type commonly found in the drinking water supplies of many large cities.

Because of their high protein and mineral content, water hyacinths also show considerable promise as an animal feed supplement. The University of Florida has successfully fed water hyacinth silage to large animals, while at NSTL we have made a water hyacinth meal by drying whole green plants to moisture contents of less than 15 per cent. This can provide a 10–20 per cent supplement to the diet of beef cattle; beyond this amount animals can suffer from a mineral imbalance due to the high levels of potassium, iron and magnesium normally found in water hyacinths.

Water hyacinth meal is a good organic fertiliser and soil conditioner, too, because of its high nitrogen and mineral content. Its high moisture retention properties would improve the condition of sandy soils, and water hyacinths can be spread directly on the ground as a mulch or compost.

Water hyacinths can also be used to produce biogas containing 60–80 per cent methane, which is a promising substitute for natural gas. Our research shows that 374 litres of biogas can be produced per kilogram of dried water hyacinth; its fuel value is 21 000 Btu per cubic metre, compared with 31 600 Btu/m³ for pure methane.

A continuous supply of water hyacinths can be grown in domestic sewage lagoons where, as we have seen, they can also perform a valuable anti-pollution function. A hectare of water hyacinths fed on sewage nutrients can yield 0.9–1.8 tonnes of dry plant material per·day. This biomass can produce 220–440 cu m of methane with a fuel value of 7–14 million Btu. In addition, the sludge that remains after fermentation is a useful fertiliser, because it retains most of the nitrogen, all of the phosphorus, and other minerals.

The clogging of waterways by water hyacinths is a serious problem in many developing countries. Although these weeds generally contain less protein than cultivated water hyacinths do, they are still very useful as a substrate for biogas production. The Sudanese Government (with the assistance of NASA through the US National Academy of Sciences) is experimenting with small-scale digesters to process the thousands of tonnes of water hyacinths mechanically harvested from the White Nile.

The water hyacinth is a warm-weather plant which flourishes in tropical and subtropical regions, but its range could possibly be extended by utilising the heat from raw sewage, by greenhouse-type canopies, or by using thermal discharges from industrial operations and power plants. The hot water from nuclear power plants is especially appealing, because the plants could act as an added safety filtration system for removing radioactive elements. At NSTL we are also experimenting with the use of duckweeds for sewage filtration during cold months when the water hyacinth is temporarily inactive.

Experiments at NASA's National Space Technology laboratories have shown, therefore, that in tropical and subtropical conditions water hyacinths have the ability to absorb organics, heavy metals, nutrients, and other chemical elements from wastewater while producing large quantities of plant material. This water hyacinth biomass, when grown in sewage free of toxic metals, is a potential source of protein fertiliser, methane gas and other valuable products.

12

Energy can be green

DR ROGER LEWIN

26 May, 1977

Biologists admit that plants are particularly profligate in their use of the Sun's energy: the maximum energy conversion of incident radiation by a green plant is a mere 5 per cent, and it is usually much closer to 1 per cent. This figure looks even more discouraging when compared with the 12–15 per cent efficiency of solar cells. But the product of photovoltaic systems is electricity which must be stored separately, a process which immediately knocks the overall efficiency to a figure below that of plants. In plants, energy conversion and storage go on in the same package. Without doubt, therefore, plants *will* be important in future energy equations, but proponents must be careful not to oversell them for fear of sowing seeds of premature disillusionment.

The gathering on this topic, organised with the enthusiastic support of the French Government in a mountain resort near Grenoble in May 1977, focused on the variety of options offered by plant systems and tried to dissect out the important factors which limit their productivity. Energy farms, power stations fuelled by "wastes", exotic algae, each had its enthusiastic proponents; so too did the notion of synthetic hydrogen-producing systems based on biological models, a crucial advantage of which is that by exploiting biological systems in a synthetic context many of the productivity-limiting barriers inherent in plants may be circumvented. One overall impression is that in the race to plug the energy gap that will be left as oil supplies are exhausted towards the end of the century, biology will not be offering one *single* technology as a replacement: there will be a matrix of technologies, each unit exploiting different aspects of plant production.

Biological systems harvest resources that, though locally variable, are virtually unlimited: sunlight, carbon dioxide, and water. Plants trap the Sun's energy to build complex chemicals from simple ones, the basis of which is a carbon backbone. When we make use of the plant material, either as a source of heat or of chemicals, we sooner or later break it down again to carbon dioxide and water: biological energy conversion systems therefore represent a *renewable* energy source, and this is an important contrast with fossil fuels (coal, oil, and natural gas). The real point, then, is how well can plants, large and small, substitute in a world built around oil as a convenient liquid fuel and chemical resource? Ultimately the answer will depend on basic biology and inventive technology.

Each year the Earth's surface receives 3×10^{24} Joules of solar radiation of which 3.1×10^{21} J are trapped by plants and converted into chemicals (stored energy): 99.9 per cent of the radiation is therefore unused. Of the energy stored by plants each year only about 0.5 per cent finishes up on our plates as food while the rest cycles more or less rapidly through the world's biomass. There therefore appears to be scope for improvement. In high technology countries people are much more hungry for energy to run their daily lives than they are for food energy on which to run their bodies. In Britain, for instance, food represents only one-fiftieth of people's total energy consumption. In spite of this imbalance no one is arguing that the best land should be given over to growing energy crops rather than food crops. Indeed, both US and European research programmes are searching for crops that will grow where currently nothing is grown systematically: for instance, ocean kelp farming is being investigated in the US, and the Irish are studying the possibility of growing willow trees in worked-out peat bogs. There is a persuasive line of argument which says that if you can grow it, eat it; or rather, biological inventiveness should be directed to new food crops rather than diverting resources to problems that can be solved by other means.

As well as attempting to develop new plants, either through more or less conventional plant breeding or by fancy genetic engineering, basic biologists are looking for ways by which to improve the efficiency of photosynthesis. Half the Sun's spectrum is useless to plants because their pigments absorb in only two main regions (400–500 nm and 600–700 nm). More light is wasted through reflection and dissipation in the leaf. Further losses occur as a result of the level of quantum efficiency and apparently wasteful respiration, bringing the theoretical maximum efficiency to around 5 per cent. Conditions in the field usually reduce this by at least half. Theoretically there is great scope for enhancing photosynthetic efficiency as well as finding ways of encouraging plants to divert higher proportions of their material resources to economically important parts of the plant (away from stalks and into seeds, for instance). And the recent discovery that many important crops are apparently energetically extremely profligate offers another opportunity for boosting photosynthetic yield: as well as manufacturing chemicals in the light many plants indulge in what is termed photo-respiration — they burn up in an apparently wasteful way at least half of their photosynthetic products. Wheat, spinach, tobacco, and hay are in this category (known as C3 plants), whereas maize, sugar cane, and sorghum are much more economical and do not photo-respire (they are termed C4 plants). There is a lot of head-scratching about ways to block at least partially the photorespiration, thus enhancing net photosynthetic productivity at a stroke.

In drawing up the energy budget in any kind of energy farming the cost of fertiliser rapidly establishes itself as a major input: the figure is often one-third of the total fossil fuel energy consumed. For this reason many people are interested in engineering nitrogen-fixing genes into plants. There are countless genetic and biochemical fences to be cleared before this attractive prospect is truly in sight, but once it is reached it will certainly impact significantly on energy budgets. One slick analysis of the energy budgets involved in farm-ing hardwoods for fuel has recently been carried out by the MITRE Corporation in Virginia. On a six year rotation using high tech-nology management Bob Inman calculates the budget to be 15 units of energy produced for every unit put in. The balance is in fact somewhat less favourable than this figure suggests because the quality of the input energy is higher than that produced: you can't equate a barrel of oil directly with a pile of logs.

There are many problems with silviculture (wood growing), not least of which is making sure that the plantation is not overrun by unwelcome invaders. But Inman's analysis shows it to be close to commercial viability, a situation that is bound to improve as the cost of oil continues to rise. The most immediate impact of biomass on energy production, however, is likely to be use of wastes. Indeed, the sugar cane industry is already showing the way. Practically all cane factories generate their own electricity (and usually an excess) by burning the pulped fibre. Incidentally, sugar cane has featured in an Australian study on energy farming: the sucrose can be used as a chemical feedstock and the fibre as an energy source. Millions of tonnes of straw are burned in farm fields every year: in Britain the figure is 4.5 million, and in the US around 200 million. In most countries there is enough straw burned to provide the farms with most of their electricity. Questions of collection and transport are of course important here, as indeed they are in all aspects of waste utilisation.

Microbiologists are also interested in energy problems and there are a number of schemes involving, for instance, growing algae in sewage. Algal culture inevitably has problems of separation and high water content (both of which imply energy expensive processes), and a more profitable exploitation of these organisms may very well be in specialist areas. One such is a remarkable alga (*Dunaliella salina*) which grows in the Dead Sea and produces large amounts of glycerol as a defence against the high osmotic pressure. Under the right conditions this creature can be persuaded to manufacture glycerol as 85 per cent of its dry weight, a property that allows algal glycerol production at commercially very favourable prices. Glycerol could serve as an important chemical feedstock, and the Israeli Government is investing $0.8 m in developing the system.

Algae sometimes figure in projects for producing hydrogen (a potential fuel itself or for making fuels) from biological systems, as too do "synthetic" plant systems. This latter approach is currently at a very early stage of research, and yet it is conceptually very appealling. The idea is to construct systems based either on totally synthetic components (modelled on the biological structures) or on a mixture of biological and synthetic components. Either way, one great advantage of the system is that many of the natural inefficiencies of plant photosynthesis could be avoided. A research plaything it may be at the moment, but in the future it could be very important.

The European Economic Community is spending $17 m over a 4-year period on solar energy. Of this 43 per cent is spent on photovoltaic research compared with 5 per cent on biomass work. The signs are that in the future biomass will gain more support. Meanwhile the US's figures, as one might expect, far outstrip those in the EEC: it is spending $12.7 m this year on biomass research. Next year the figure is due to rise to $20 m, and the following year it will be at least $40 m. Biological solar energy conversion systems have arrived!

PART THREE

Nature's oil

13

Put a sunflower in your tank

DAVID HALL
26 February, 1981

Plant power produces one particularly attractive commodity – natural oils. These need little processing, and research has shown that engines can run on them. The day of home-grown diesel fuel may not be far off.

On the shelves of your local supermarket you will find oils extracted from plants – peanuts, sunflowers, maize, soya beans and olives; all are common sources of "vegetable" oils. Concern about health and convenience have encouraged the consumption of vegetable oils but it may not be long before these oils are also accepted as fuels for transport. In some parts of the world the prospects for this are very good. At the moment these oils would be too expensive as fuels. However, experiments with "organic" fuels in Brazil, South Africa, Zimbabwe, the United States, Austria, Australia, Japan and undoubtedly a number of other countries, have shown that vegetable oils do work. They can be used either pure or blended with other liquid fuels, such as ordinary diesel fuel or alcohol, to run diesel engines.

Plant oil is, therefore, a promising alternative to alcohol when it comes to turning plants and organic materials into useful fuels. Vegetable oils have the advantage that it is easier to extract the fuel – all you have to do is to squeeze the oil out of the plants. And recently there has been something of a boom in research and development projects to study vegetable oils as "renewable" fuels.

Rudolf Diesel would not be too surprised at this great interest in alternative fuels for his engines. In 1911 he wrote: "The diesel engine can be fed with vegetable oils and would help considerably in the development of agriculture of the countries which will use it. This may appear a futuristic dream but I can predict with great conviction that this use of the diesel engine may in the future be of great importance." Diesel tried all sorts of fuels and concluded that practically anything could fuel a diesel engine if it could be injected into a cylinder and would ignite at the temperature generated by high compression. Since his work, any manufacturer of diesel engines that wants publicity has run an engine on some exotic fuel.

Sunflower oil and peanut oil, for example, release something like 90 per cent of the energy released by conventional diesel oil. However, the fuel efficiency of these oils is only about 4 per cent less than that of diesel, and if the "crude" vegetable oil is processed these vegetable oils can actually out-perform diesel oil, but more of that later.

In the Second World War, China developed an industrial process for cracking vegetable oils (breaking them down into smaller, more volatile molecules), mostly oil from tung nuts, and turning them into motor fuels. The Chinese also used rapeseed oil and peanut oil. They had previously found that diesel engines could burn vegetable oils, but these are more viscous than conventional diesel fuels and have to be injected into an engine at a higher pressure. Another disadvantage is that vegetable oils produce more carbon when they burn; tung oil in particular can also form large molecules which literally gum up engines. So the Chinese opted for the so-called vegetable-gasoline programme in which the vegetable oils were processed so that they could fuel an engine that did not have to be modified to cope with the fuel's oddities.

Vegetable oils can be extracted from an impressive range of plants. Besides sunflowers, palm, olives and the better known plants, there are more exotic species such as jojoba and guayule – both desert plants – castor bean, rapeseed, milkweeds, eucalyptus, squashes, copaiba, malmeleiro, babassu nut and so on. Oils from these plants, and almost certainly from others, are being investigated in many countries. But before we can turn confidently to these oils as fuels we need to know how much they cost as well as how much energy we have to invest in growing and processing the plants. There is no point in growing plants to produce vegetable oil for fuel in 2 litres of premium fuel are consumed in the process of "growing" 1 litre of oil. Here the key factor is the productivity – how much oil can farmers produce on a hectare of land – and the cost and ease of production. Trials in South Africa and the Midwest of the US with sunflowers and soya beans show that yields can be 1 tonne of oil per hectare with unsophisticated processes for extracting oil (something like 40–50 per cent of weight of the harvested crop is oil) while yields as high as 2 tonnes per hectare are not uncommon, and claims of up to 5 tonnes/ha have been reported. Oil produced in this way costs about $2 a gallon (20p a litre) and the ratio of energy produced in comparison with that put into growing and processing the crop varies between 3 and 10 to 1.

Energy ratios have always been controversial when considering biomass energy. When crops are grown and fermented to make alcohol, the benefits are even less obvious – energy ratios are lower than they are for vegetable oils but alcohols are more versatile as fuels. The returns from producing vegetable oils are such that, according to some estimates, if a maize farmer were to grow sunflowers on 10 per cent or so of his land the oil from the crop would make him self-sufficient in fuel.

One of the great benefits of growing plants for oil is the wide variety of species available. We can choose whichever plant is best for the growing conditions in a particular area. Sunflowers are good for many parts of the world – from South Africa to the Soviet Union and the US – but other vegetable oils may be more suitable for other countries. Only extensive research and trials will make it possible to choose the right plant for an area.

Brazil's experience – for example, its Proalcohol programme for extracting ethanol from crops (see "The gasohol gamble", p. 165) – is a good indicator of governments' interest in biomass. In 1980, Brazil began to investigate a diverse range of biomass feedstocks for fuel production. Researchers are testing oils extracted from soyabean, sunflower, peanut, rapeseed, dende palm and castor, along with oils from indigenous plant species such as "pinhao" (*Jatropha*), *Copaifera*, *Croton* and *Tolouma*. They are also testing "straight" oil as well as various blends with ordinary diesel oil. Numerous large-scale trials are under way, including the conversion of the bus fleet in Brasilia, the country's capital. The National Institute of Technology in Rio de Janeiro completed a series of tests in January, running local buses on mixtures of diesel and vegetable oil. In one trial the fuel was a mixture of 80 per cent diesel and 20 per cent peanut oil, and the other trial used 73 per cent diesel, 20 per cent palm oil and 7 per cent ethanol. In this trial the bus ran about 500 hours for 10 000 km (it had previously clocked up 60 000 km on diesel fuel alone). After the trial the bus reverted to normal running on diesel oil, without any readjustments. The trial showed that with the mixture the fuel consumption was 3.4 per cent less than with diesel fuel alone. The plan is that this year vegetable oils will meet 6 per cent of Brazil's demand for diesel fuel, rising to 16 per cent by 1985.

The Brazilians are among those looking closely at the economics of palm oil as a substitute for diesel oil or as an "extender" that can be added to make the petroleum-based fuel last longer. Many tropical countries already have an established industry based on the African oil palm as a source of vegetable oil. Yields of 4 tonnes of oil/hectare/year have been claimed – this exceeds the yields of alcohol from sugar cane and cassava, which are around 2–3 tonnes/ha/year.

There are some advantages in making fuel from palm oil rather than making ethanol from sugar and starch crops. It is easier to extract the fuel from palms, year-round production is possible by

Palm oil nut, Malaysia.

D

Oil palms in Malaysia: tomorrow's diesel fuel?

continuous harvesting, there are not so many polluting byproducts, and one of the "leftovers" from the process is a protein-rich animal food. On top of that, growers can use the whole plant to make a number of products, the energy return is very good, and water is not needed during processing of the fuel.

The number of different plants that can yield oil is legion; and different varieties are suitable for different parts of the world. For example, in the Pacific islands of the Philippines, Fiji, Samoa and Papua New Guinea, a collaborative R&D programme led by Australia is investigating the possibility of using coconut oil. In Australia itself there is growing interest in peanut, soyabean, sunflower and rapeseed oils.

South Africa's search for substitutes for imported oil and indigenous coal – these provide nearly all the country's energy – has prompted it to look at vegetable oils. The Division of Agricultural Engineering, in Pretoria, has fuelled diesel tractors with sunflower oil. And in neighbouring Zimbabwe, an agricultural engineering group has studied sunflower oil, used both "neat" and mixed with diesel fuel. The South African researchers compared the properties – calorific values, combustion efficiencies, octane numbers and so on – of various grades of sunflower oil, ranging from oil that has not been processed after extraction to oil "degummed" to remove some of the contaminants and refined oils. Nine tractors were tested, with engines run continuously for up to 58 days on a fuel containing 20 per cent sunflower oil. These engines operated normally, with only minor deposits in the combustion chambers, cylinders and piston-ring grooves. It turned out to be easy enough to start engines fuelled with 100 per cent sunflower oil. And there were no adverse effects after 100 hours running at maximum power. One tractor subsequently completed more than 1000 hours of trouble-free operation on a farm.

The engineers in Pretoria did find that when they fuelled diesel engines with a high percentage of sunflower oil, the injection nozzles became clogged, especially when the engines were not running at full power. But they overcame this difficulty by adding a different fuel filter, and by tuning the engines. They also found that contaminants from the fuel – mostly partially burnt oil – found their way into the engines' lubricating oil, but only if the oil was not changed every 200 hours.

The South African research group has decided to process sunflower oil in an attempt to eliminate these problems. The idea is to esterify the oil to ethyl or methyl esters using ethanol or methanol.

(This is done by chemically combining an acid and an alcohol to form an ester by removal of a water molecule.) It is easy enough to do this; all that is needed is to add acid to the oil and to heat it to 30–40°C for a few hours. The fuel mixture produced in this way consists of fatty-acid ethyl esters, some unreacted oil and ethanol. The distillation and viscosity properties of the fuel produced in this way are very similar to those of ordinary diesel fuel, but the mixture causes less coking and exhaust smoke than diesel fuel; it also has a higher thermal efficiency than diesel fuel. In other words this sunflower-oil/ester mixture can actually improve performance.

Researchers in Brazil have carried out similar projects with processed vegetable oil. They found that it is better to esterify with methanol because it is easier to separate the methyl ester after settling. On the other hand, the ethyl ester has to be distilled to obtain a pure product.

In South Africa work continues on the long-term effects of sunflower oil as an additive to diesel fuel. Researchers there are looking into such aspects as the modification of fuel-injection equipment, the use of additives to the fuel and to lubricating oils, the design of oil extractors for farms and cooperatives, storage and refining of the oil. South African farmers are fortunate in having over half a million hectares of sunflowers in their country – three times the area of a decade ago – and extensive trials and plant-selection programmes have increased yields by some 2½ times since 1970 to an average of 1.2 tonnes per hectare in 1977. In fact, with new hybrids some farmers have achieved yields as high as 4 tonnes/ha, and the seeds produced contain 45–50 per cent of their weight as soil.

Naturally enough, a new energy alternative with such promise has aroused interest in the United States. The Department of Agriculture, the North Dakota State Extension Services and the American Soyabean Association are among the groups that have studied vegetable oils as fuels. Researchers have tested various mixtures of oil and diesel fuel. One series of trials showed that a mixture containing 75 per cent sunflower oil had a very similar efficiency to that of straight diesel fuel. Here too engineers are investigating the optimum operating conditions, the use of additives, esterification, filtering, degumming, and so on. They are also working on the extraction and processing of the oils to minimise costs and to simplify the process.

To the American farmer there is a considerable attraction in being self-sufficient in fuel. To begin with, it will often be easier for farmers to extract oil from plants than it will be for them to ferment starch and sugar crops to make methanol. Naturally enough, one concern is the cost of vegetable oils. Unfortunately it is difficult to make sensible price comparisons because there is a tax exemption of 12 cents a gallon on diesel fuel for farmers in the US – they now pay about $1.37 a US gallon (36 cents a litre). There is no such tax advantage for vegetable oils. However, calculations show that on a heat of combustion basis, oil priced at 25 cents a pound (US) is equivalent to diesel fuel at $2.09 a gallon: alternatively, diesel fuel at $1.13 per gallon is equivalent to soya oil costing 14 cents a pound.

Farmers in the US have achieved yields of sunflower seeds between 500 and 3000 lb/acre (0.6–3.5 t/ha) within three months of planting. The seeds can be grown commercially in all the agricultural areas of North America, often on poorer land and with minimal amounts of water. Oil yields vary between 0.3 and 1.6 t/h – about 100 gallons or 800 pounds of oil per acre with a seed yield of 2000 lb/acre (2.3 t/ha), which works out at between 0.7 and 4.4 barrels of oil per acre if you prefer these units. In the southern states a plot of land might be able to produce two or even three crops of sunflower seeds a year. Clearly, the US has to conduct research before sunflower oil can become an accepted fuel, but this looks like a "sleeper" in the R&D programme for "synthetic fuels".

Even in Europe there is quickening interest in rapeseed oil as a result of work at Wieselburg in Austria. There rape, sunflower, and soyabean oils have been tested as tractor fuel in 1000-hour runs. This work continues; and in other European countries researchers are collaborating with engine manufacturers to test plant oils.

Two other plant oils have also been looked at as fuels. In the region around the Mediterranean, olive oil production exceeds demand, and the soil would make a good substitute for diesel oil. The second possibility is eucalyptus oil, which can be extracted from the leaves of this tree – something that the pharmaceutical industry has been doing for years to make eucalyptol for toothpastes and proprietary medicines, for example. As yet the oil yield from eucalyptus leaves has been low, about 1 per cent, but recently there have been claims that selected species could increase this to 25 per cent. And work in Japan has shown that eucalyptus oil can be used straight, or in blends of 70 per cent oil in gasoline. One drawback to eucalyptus oil seems to be that it is hard to start a cold engine, but the power output is said to be better, and there is less carbon monoxide pollution than with gasoline.

So far I have focused on the positive aspects of growing plants for their oil. But as with all biomass-for-fuel programmes, we have to recognise the problems. No fuel system is perfect, including vegetable oils. Plants need land and compete for other resources that are required in agriculture: thus they compete with food production. Each country has to make its own decisions on priorities when it comes to a competition between food and fuel. However, crude oil is expensive and the cost is not likely to fall. Vegetable oils can be grown locally, and they are renewable. The day of the sweet-smelling diesel bus and truck may not be all that far off if the cost of crude oil continues to climb and if we can produce vegetable oils in sufficient quantity and at the right price.

PART FOUR

Wood – too precious to burn?

Wood is the best known and most common biomass fuel. It is also the most threatened. While forests still cover about one-fifth of the world's land, and open woodlands another 12 per cent, the resource is being rapidly depleted in areas where people depend on it for fuel. As Keith Openshaw explains in the first article in this Part, "Wood fuels the developing world", four out of five families in the developing world depend on wood as a primary source of energy. In some countries 99 per cent of the population burns wood. Where the population is dense, for instance in the Indian subcontinent and parts of Africa, forests do not get a chance to regenerate themselves. Industrial forestry, which usually means taking timber for export to the developed world, makes the problem worse.

The people who suffer, as Anil Agarwal's "Forgotten energy crisis" explains, are the poor people who need wood for their cooking stoves. Many Indian women walk 10 km a day, 5 days a week, carrying 25 kg of wood on their heads. This is the real energy crisis: and in many parts of the world the situation is deteriorating. For example, Nepal is consuming its forests seven times faster than it is replacing them, and Kenya is likely to face a severe shortage of woodfuel by the 1990s.

The problem is one that Britain faced two centuries ago, and overcame by turning to fossil fuels. But importing fossil fuels is a particular burden for the very same countries that are at risk of destroying their forests. Kenya spends a quarter of its income every year on importing oil. The solution must lie in managing forests carefully, and burning the fuel efficiently.

It is clear by now that large-scale biomass projects have just as much effect on the environment as other energy technologies do. As we shall see, the effects of devoting huge areas to sugar cane in Brazil to make fuel alcohol, or laying down vast acreages of sunflower or palm for fuel oil, are no less than those of any other form of intensive agriculture.

But neither can we assume that the apparently innocuous gathering of sticks or collection of cow-dung, activities that provide most of the Third World's fuel, are necessarily benign. Even when people gather sticks without damaging the trees (a big proviso), removing dead wood robs the environment of food for invertebrates, and when invertebrates are reduced so too is bird life. Burning dung robs the soil of potential fertility: each tonne burned means a potential loss of 50 kg of grain.

However, it is at least theoretically possible to produce biomass fuel in ways that benefit wildlife, just as it is theoretically possible to grow food or commercial timber without necessarily destroying animals and wild plants. Thus, Britain's wildlife suffers grievously from the removal of hedges; but if the trimmings of hedges were garnered specifically for fuel, this would provide some economic excuse for keeping them.

Similarly, a highly productive way to take low-grade timber from trees is to pollard them – cut them off at ground level or a little above, and allow the remaining stumps, with their deep and extensive roots, to throw up a never-ending succession of new, fast-growing branches.

Pollarded woods, of trees such as chestnut and ash, have at times formed the basis of important European economies, producing fencing and fuel. They also provide some of the richest ecosystems to be found in temperate climates, tending as they do to combine the flora and fauna of open grassland with that typical of forest, and maintained in a state of constant ecological tension by the activities of woodsmen.

It is often said that Britain's farmland is too valuable to be used exclusively for producing food; it should also serve as parkland for people to enjoy, and as reserves for wildlife. Farming, wildlife and leisure can be combined, at least in theory. So it is possible, at least in countries such as Britain, to see aesthetic considerations tempering to some extent the crude provision of biomass fuel. It is a possibility that must be taken into consideration whether considering Britain's craze for woodstoves (which *New Scientist* described as "trendy pollutors") (p. 108), or by energy planners in the Third World. At the moment, they have an uphill struggle persuading people to spare a thought for wildlife. As Anil Agarwal comments (p. 90), people who have to scrounge around for plant wastes for their stoves do not worry about the environment.

14

Wood fuels the developing world

KEITH OPENSHAW
31 January, 1974

The energy crisis in Third World countries is at least as serious as the much publicised problems of the developed nations. Most of the population in the developing countries are peasants living in a subsistence economy. Their energy needs for cooking, heating, protection (a fire keeps away unwanted insects and wild animals), and cottage industries are met almost entirely by burning wood which is collected from the surrounding countryside. Their urban counterparts also depend on woodfuel, mainly in the form of charcoal, for their everyday energy needs.

Over 95 per cent of households in developing countries, where woodfuel is readily available, use it as a primary source of energy. This falls to about 80 per cent when all developing countries, including wood-poor regions and desert countries, are counted. Likewise industry in these countries such as brick and ceramic manufacture, fish and tobacco curing, cassava meal, rubber, sugar and wattle tanning production, restaurants, tin smelting, and steel industries all burn woodfuel to some degree.

In areas where a lot of wood is available, people collect only dead branchwood. But where shortages occur people take young saplings. Eventually mature trees are felled, especially by the "urban" charcoal producer who needs large quantities of raw material within an economic transport distance of the urban centres.

The demand on woodlands and forests has denuded the land in certain areas. In consequence the population has either to move on or make do with substitutes such as cattle dung, to the detriment of the soil. The removal of the vegetation cover has also increased soil erosion which, at least, causes loss of topsoil and, at worst, results in massive flooding in river valleys and delta regions. The principal cause of the recent floods in the Indian subcontinent was the

Families in the Third World spend much of their time and energy collecting wood for cooking. These burdens represent a day's work for these boys in drought-ridden Northern Ethiopia.

removal of tree cover in the catchment areas for fuelwood (and shifting cultivation). Likewise, the continual cutting of fuelwood in the Southern Sahara/Sahel region of West Africa has facilitated the rapid shift southwards of the desert, and has made living in this area even more precarious.

For the past five years the UN's Food and Agriculture Organisation has undertaken surveys of wood consumption in East and West Africa and in South-East Asia. The main emphasis of these surveys was to determine the consumption of industrial wood – sawnwood, panel products, and paper – and to forecast future requirements so that meaningful plans could be made for the development of the forest and forest industries. Nevertheless, all forest products were considered and in each country the present and impending shortages of woodfuel in the different regions was the most urgent problem.

These surveys indicate, if projected to all the developing countries, that although woodfuel enters the industrial process in a minor way, it accounts for over 85 per cent or about 2000 million tonnes (air dry) of total annual wood consumption. Indeed for the world as a whole woodfuel represents about 65 per cent (2300 million tonnes) of 1972 wood use. This wood is equivalent to burning 1300 million tonnes of coal.

On a per capita basis, average consumption in developing countries is just over 1 tonne per year but this increases to 1.3 tonnes/year, or 8 tonnes per family per year when only those who use wood are counted. As the wealth of the population increases there is a switch to fuels other than wood, especially in urban areas. Therefore average per capita consumption is decreasing slightly but in absolute terms population increase is increasing total demand at a rate slightly less than the population growth of the developing world, or about 2.5 per cent per year. However, because of existing shortages, consumption is increasing at a somewhat slower rate than the projected growth in demand. In other words there is an excess in demand over supply which will continue and widen if nothing is done to correct it.

In practical terms this shows itself in a number of ways: ever increasing time spent on collecting fuel-wood; price increases; and the cutting of living trees (first saplings and finally mature trees), leading eventually to the two extremes – an increase in desert areas or calamitous floods.

The shortfall of supply over demand leads to a reduction in area and volume of forests and woodlands. While a temporary decrease in forest capital can be made good, a continual decrease will lead to disaster not only for the developing countries but for the whole world. Forests and woodlands are a considerable source of oxygen and the developing world's woodlands make up half the total forest area.

Woodfuel consumption in some developing countries

Country	1973 GDP per capita £	Percentage of GDP derived from subsistence sector	Consumption of woodfuel per capita/yr tonnes	Woodfuel consumption out of total timber consumption %	Population using woodfuel %	Charcoal consumption out of total woodfuel consumption %	Urban population %
Tanzania	40	30	1.8	96	99	4	7
Gambia	50	25	1.2	94	99	26	23
Thailand	80	20	1.1	76	97	45	15

With proper planning, the shortfalls in supply could easily be reversed. Fuelwood plantations could be grown on a 6 to 10-year rotation planting such species as eucalyptus, Indian neem and gmelina, giving an average annual yield of about 16 tonnes per hectare (including branches) – enough for an average-sized family. As plantations are far more productive than the forests and woodlands, the latter could be cleared for other uses (such as plantations) without sacrifice. Such plantations if grown for the non-subsistence sector of the community could yield about 5 per cent per year on invested money, assuming a selling price of 80 pence a tonne for the standing tree. It would also provide many rural people with a cash income from plantation work and the manufacture of charcoal.

However, fast-growing plantations could also meet some of the energy requirements of the Western world. We could not only rediscover the use of fuelwood, charcoal, and the by-products of charcoal manufacture, but also use wood fibre, a renewable resource, as a raw material for the petroleum and allied industries. Already work is being undertaken, notably in the US, and a pilot plant has been established to produce petrol with a low sulphur content.

Whether such a plant is a commercial proposition depends very much on the price of crude oil. With current technology a tonne of woodfuel can produce two barrels of oil. At an oil price of $10/barrel this wood-to-oil process becomes competitive. Because of favourable climatic conditions and relatively low labour costs, many areas of the developing world could be in a position to enter this market with large-scale short-rotation forest plantations.

However, the immediate need of these countries is to solve their own energy crises, and by and large they have not enough trained people or resources to undertake massive plantation schemes. "Soft" loans, especially when they are not tied to capital-intensive equipment, may be hard to obtain, particularly with today's high interest rates, but they can be justified on ecological grounds, rural employment, and long-term economic terms. Therefore, it is not only essential but also in the interest of mankind that the rich nations provide trained personnel, aid and loans to help solve this problem. The developing countries could then play a greater role in meeting the energy needs of the whole world instead of being in a position where they have insufficient resources to supply their own requirements.

15

The forgotten energy crisis

ANIL AGARWAL
10 February, 1983

It is easy to forget that most of the energy that people in the poor world consume goes to cooking food. But to ignore this human need invites a catastrophe.

Even though India is the world's tenth largest industrial power, half of the energy its people consume, excluding animal energy, is spent on cooking food. This is nearly twice what agriculture and industry consume. In many under-developed African countries, cooking may take three-quarters of all energy. In most developed countries, cooking accounts for less than 5 per cent of national energy consumption.

The amount the poor world consumes for cooking is high in absolute as well as in relative terms. A Third World family of six would commonly burn about 3.6 tonnes of wood per year; nearly 12 times the average that a British family consumes. The wood-stoves that cook the food are extremely inefficient. However, energy planners in the Third World tend to overlook this essential need even though it is the most crucial for human survival and the most sensitive to environmental conditions.

For India, despite the massive development of oil and coal, non-commercial sources of energy, such as firewood, agricultural wastes and cow dung, are crucial. In 1975–76, 133 million tonnes of firewood provided India with 28 per cent of its total energy. Another 73 million tonnes of dung and 41 million tonnes of agricultural wastes are burned each year. These non-commercial sources provide as much as 87 per cent of India's cooking energy. Coal and kerosene contributed only 8 per cent of the total cooking energy in 1953–54 and rose to a peak of 14 per cent in 1970–71. After the rise in oil prices, this dropped to 13 per cent in 1975–76. It is a fallacy, although a common one, that firewood, dung and agricultural wastes are burned mainly in the villages. As much as 56 per cent of the domestic energy consumed in cities comes from non-commercial sources. According to India's planning commission, nearly three-quarters of households in towns and cities depend on non-commercial energy for cooking.

Firewood provides half the cooking energy in Indian cities and 70 per cent in villages. The price of coal and kerosene is rising rapidly, and the forests are being denuded. People will find it progressively difficult to meet their basic need for cooking energy. As a result, the price of both commercial and so-called non-commercial fuels is rising rapidly. Between 1970–71 and 1980–81, per capita income rose by 143 per cent and the wholesale prices of all commodities rose by 150 per cent. But the wholesale prices of commercial fuels – power, light and lubricants – rose by 264 per cent. In many cities the price of firewood has nearly doubled in the past 6 years. The demand for firewood now far outstrips supply. Populous Uttar Pradesh, for instance, produces only 18 million cubic metres per year. By 2000 the demand will be 63 million cubic metres per year.

Experts know little about how people are coping, but research in the past couple of years shows that for the majority of India's people the cooking energy crisis is already here.

In hills and plains, women and children spend hours travelling long distances to collect fuel. In Dwing, a village in the high Garhwal Himalayas of Uttar Pradesh, women walk at least 10 kilometres three out of four days, for an average of seven hours per day, to bring back about 25 kg of wood with each headload.

Even rich villages have their energy-poor. In Navli, a prosperous village close to the famous cooperative Amul Dairy in Gujarat, irrigation, cash-crop farming and animal husbandry have made several families very rich. But with firewood supplies dwindling, and prices of imported firewood rising, farmers with large numbers of cattle have set up biogas plants, and large landowners have started collecting their crop residues for fuel. So even though more cow dung and crop-residues are available, people who have no land face increasing hardships. They are now totally dependent on the landlords. As one poor woman in the village laments: "There isn't enough money for food, so where is the question of buying firewood? We sometimes uproot small plants from around fences. When we are caught, we are fined. We have lost our livelihood and, it seems, our lives too."

To what extent these isolated scenes of rural India add up to a true picture of the situation across the country is difficult to say. Studies of India's energy economy are rare, and knowledge of trends is even rarer. But we must assume that many millions of poor rural women are caught in a vicious cycle: they eat food to get human energy, and then spend all this energy in producing food and collecting the energy needed to cook it. There is little else left in their lives.

We know even less about supplies of cooking energy in urban India, especially in small towns. Nearly three-quarters of firewood and about half of the dung in urban India is bought. A substantial part of this firewood is brought in from the villages to be sold in the cities, although nobody really knows how much. Thousands of people carry firewood on their heads to Ranchi every day from nearby villages. All the three daytime trains from Dhronachalam to Kurnool in Andhra Pradesh bring with them a garland of firewood bundles strung outside carriage windows. In Kurnool, these trains are known as "firewood specials".

The Indian Institute of Science in Bangalore has shown that the city draws in nearly half a million tonnes of firewood every year. Most of it comes on diesel trucks. Considering that most wood-stoves have an efficiency of only 5–10 per cent, for every 100 calories of energy provided by firewood for cooking, 25 calories are spent in the form of imported petroleum to transport this firewood to the city.

Although the government has been pushing liquid petroleum gas (LPG) in metropolitan cities, only the upper middle class can afford it and the demand far outstrips the supply. In towns, kerosene is the main fuel for cooking. It is now almost certain that more people in Indian cities will become dependent on kerosene for cooking in the near future. Bombay's population, 1 per cent of the national population, burns one-tenth of the kerosene consumed in the country. In 1981 the government had to resort to crash purchases of kerosene to meet shortages.

In fact, the erratic supplies of kerosene and LPG make it impossible for many middle class families to rely on one source of energy. In Calcutta, where the electricity supply is erratic, it can take a month to replace an exhausted LPG cylinder. Because the supply of kerosene is not assured, many families turn to a fourth option: coal. But as new houses generally do not have chimneys, burning "angithis" (coal stoves) means a lot of smoke. Many landlords in Calcutta, for example, make prospective tenants promise not to burn coal. Anyway, few families can afford all four forms of cooker.

There is still no study that describes what families at different levels of earning spend on energy for cooking, and how this changes over time. In cities, poor families spend a much larger proportion on energy – often as much as 12–15 per cent of their income. For a relatively prosperous family, spending on cooking probably drops to less than 5 per cent of the income. Considering that poor families ought to spend about 80–90 per cent of their income on food, rising fuel prices mean that they must be eating less. For an already malnourished population, this is unfortunate. Reports from Bangladesh, Nepal and Pakistan show that some rural families have to spend a quarter of their incomes on fuel.

The shortage of cooking energy not only causes problems with nutrition, but spreads disease as well. The Food and Agriculture Organisation's document *Agriculture: Toward 2000*, points out that with the production of fuelwood expected to fall short by 40 per cent, "many poor people will not be able to cook their food adequately. This can have serious nutritional and health consequences. The digestibility of food will decrease, and also the incidence of parasites ingested with insufficiently cooked meat will rise. There are reports of this happening already."

The already-widespread incidence of scabies in India can certainly be attributed to inadequate supplies of water and fuel. If people cannot heat water in winter, they will not wash. The result: scabies.

So how are the poor trying to meet the energy crunch? Who, for example, uses firewood? There is now definite evidence that villagers are trying to meet their energy needs by substituting one resource for another – agricultural wastes for firewood, for example. Contrary to popular belief, recent studies reveal that wood (especially firewood in the shape of logs rather than small twigs and branches) has become so scarce that in many villages only the rich can afford to buy it. The bulk of wood that people burn is in the form of twigs and branches. The poor rarely fell trees: they lop off small branches, twigs, roots and dead wood.

In Pura, a village in Karnataka, Professor Amulya Reddy and his colleagues from the Indian Institute of Science found that 96 per cent of the wood that poor families burned was in the form of twigs. The rest was small branches. In contrast, rich households, which get wood from their own land, depend on logs. They consume nearly one-fifth of the village's firewood. Thus Reddy concludes that gathered firewood, upon which three-quarters of the people of Pura depend, does not contribute to the destruction of forests.

There may even be a male–female dichotomy on the issue. Among the poor, it is women and children who collect firewood, and they seldom cut down trees. The men, who are interested in cash, are more likely to go for the big wood.

It is the commercialisation of fuel that poses the greatest threat to India's forests. In the cities, even traditionally non-commerical fuels are going onto the market. The same happens in the rural areas, although at a slower pace. And it is the relatively rich, in small towns and on the fringes of big cities, who create this demand. The headloaders of Ranchi and the "firewood specials" of Kurnool are responses to this. In the process the landless poor, whose numbers grow every day, try to earn a pittance.

The pressure on trees differs between regions. Families in the hills not only consume 25 per cent more energy than those on the plains, they also consume more firewood. In desert and hilly regions, where agricultural production is low, people obtain two-thirds of their energy from firewood. In the plains, its contribution drops to 37 per cent, and that of crop residues goes up to 31 per cent. Thus firewood is the most important source of energy in desert areas, where vegetation is sparse. The situation is ecologically disastrous.

The rising number of landless and marginal farmers poses a serious challenge. In 1977, 24 million households, with a total population of 114 million, each had less than 0.4 hectares of land. With supplies of firewood declining, these people will turn more to cow dung, and become more dependent on landowners. Any trend toward the commercialisation of fuels such as dung and plant residues, or the introduction of technology such as biogas plants to allow land-owning families to exploit these resources themselves, will inevitably hit the landless poor. Certainly they could easily find themselves short of energy amid plenty.

While much has been written about the concentration of land and cattle-holdings, there is little information about who owns trees. A study from a village in Bangladesh shows that 16 per cent of its families owned 55 per cent of the cropped land, and 46 per cent of the cattle. They owned 80 per cent of the trees, and took a keen interest in protecting them.

The old feudal relationship, in which a landlord took care of his agricultural labourers, is rapidly disappearing under the onslaught of the cash economy. Many farmers now leave the village poor to fend for themselves. Thus, even if biomass energy increases, the poorest villagers may find that in periods when employment is scarce there is no livelihood better than to collect some wood and sell it. People who have to scrounge around by the roadside for plant waste for their stoves do not worry about the environment.

In Gujarat, two industrial consultants recently investigated why a process developed by a Cotton Technology Research Laboratory in Bombay to make boards from cotton stalks had not been adopted, although wood was expensive. They visited three tribal villages where cotton stalks are a major crop residue: Goral, Kanpur and Choriwad. They found that, when wood became expensive, the poor villagers gathered cotton stalks as fuel. One farmer admitted that he had been *selling* cotton stalks to agricultural labourers, a new trend in the region.

Dr B. S. Pathak, of the Punjab Agricultural University, has studied the potential of agricultural wastes in a village in the district of Ludhiana. Bhundri is a relatively rich village, where several farms have tractors, and all its farmers practise "green revolution" agriculture. Although the village is rich, cooking accounts for 98 per cent of the energy its householders consume. They burn nearly three-quarters of their animal wastes as fuel.

Pathak's calculations yield an amazing result. After the village has met all its needs for fodder, the remaining crop wastes and animal wastes contain enough energy to fuel the village (for cooking, pumping water, making fertilisers) – and still leave a surplus.

But will the increased availability of energy in the village really give more energy to the very poor? There is no evidence to believe that this would be so. Even in fuel-rich Bhundri, three families have to collect plants to burn as fuel.

If a government in the Third World wants to meet the energy needs of its people without destroying the environment or creating human misery, it has to adopt an integrated national policy on energy for cooking. To date, rural planners have virtually ignored the need for energy to cook food. The few attempts to increase supplies have been marked by *ad hoc* efforts based on technology, with biogas plants, solar cookers or fuelwood plantations promoted as universal solutions.

The pattern of the supply and demand for cooking energy is complex. People live in different ecological regions, different types of settlements, and have different levels of income. Attempts to introduce new cooking-energy technologies will have to be far more systematic. Each source of energy or technology has its own economic, social and ecological niche, and it is only in that niche that it can prosper. Ever-popular blanket solutions, such as biogas plants for all and sundry, will not work.

Identifying these niches can bring surprises. For instance solar cookers, instead of becoming the answer to the needs of poor villagers and thus easing the pressure on firewood, could become a tool for the middle classes. There, they would reduce the need for oil products such as kerosene and petroleum gas. Not only can the middle classes afford solar cookers, they are also the people that the rising price of petroleum products hits the most. They may, therefore, be more keen than the fuel-gathering poor to save energy. A government supermarket in New Delhi has been selling 10 solar cookers every day for the past few months at about £20 each. Shops in Gujarat have already sold more than 7000 solar cookers. Middle class families get their investment back in 2–3 years if they cook half their lunches in the cooker.

A national policy on cooking energy also has to take into account the relationship between the energy needs of the urban and rural areas. It must also make full use of all sources of energy.

India has launched many social forestry programmes with the professed intention of meeting community needs such as firewood for cooking. But in reality, none of these programmes is helping to increase the production of firewood. The increased supply goes to those who can pay – the paper industry and the urban consumer. To make it profitable to grow wood, the state of Gujarat, where thousands of farmers have turned over good irrigated land to tree farming, has even proposed building generating stations fuelled with wood. But if this were to happen, only the rich would benefit. Thus, a second green revolution may be in the offing in which big energy production increases but the energy-poor still starve.

Gerald Leach of the International Institute for Environment and Development in London points out that, worldwide, there is no real shortage of oil products for basic human needs. But each country has to look at its own resources and solutions. India, at least, despite its large population, is not short of resources: wood, biogas, animal waste and plants.

Energy for cooking is clearly one of the biggest human needs, second only to food. The problem in meeting this need is not a lack of resources, technologies or knowledge, but of political will and organisation. It is a case of huge inequalities, between and within nations, and of starvation amid plenty.

16

Saving Nepal's forests

FABIAN ACKER

9 April, 1981

Nepal is consuming its forests seven times faster than it is replacing them. By turning to water power and making gas from the dung of its six million cows, its people could reverse the trend.

Nepal's greatest natural assets – its forests – are disappearing at a rate that could be disastrous for the land-locked kingdom, sandwiched between Tibet and India. Ironically, the forests could be saved by the very forces that are destroying them: water and cows.

Wood supplies 85 per cent of Nepal's energy. Each citizen consumes an average of 600 kg a year. But the country's forests grow only around 80 kg per person a year. The Nepalese also use wood for building, as it is far cheaper than importing cement overland.

The forests belonged to the local communities until they were nationalised in 1956. By 1968 the government realised that nationalisation had caused a decline in the area of forests, and passed laws to restore in part the original pattern of ownership and management. But, as one foreign expert said, "since then [nationalisation] a generation has grown up which knows nothing about woodland management, and it will have to be re-educated". This may be part of the answer, but the relentless statistics of population growth (2.5 per cent every year) – and the extraordinary number of cows in the country – also put pressure on the forests. The Nepalese do not kill cattle, for religious reasons, so cows consume vegetation on land that could grow food for humans, and often eat the branches of young trees. There is one farm animal for each of the 13 million inhabitants of Nepal, and about half the animals are cows.

If the area of forest falls below a certain level it could decline irreversibly, with serious consequences. One of the most obvious of these is erosion. During the monsoon, rivers carry as much as 50 000 times the water as they do in the dry season, and thousands of tonnes of silt are washed into the waterways. This not only reduces the fertility and stability of agricultural land, but also makes it less practical to use dams to control water or to generate electricity. In India, which has a similar geology, reservoirs are silting up at three times the expected rate, seriously reducing their operating life. Nepal's silt problem is as bad as India's, and erosion caused by deforestation will make it worse.

As wood becomes more scarce, Nepalese villagers have to use more time and effort in gathering it. Some villagers have already moved so as to be nearer the forest, having consumed all the wood within a reasonable walking distance. The more time that villagers spend finding fuel, the less time they have to spend on growing or gathering food. In an economy already close to the breadline, this could start a spiralling decline that could end in disaster. Yet the same energy that sweeps the rich soil into the rivers and streams could generate electricity to replace energy derived from burning wood. And the cows that are now such a burden on the Nepalese could also alleviate the problem. Already more than 200 biogas plants fuelled on cow dung are replacing wood and paraffin for cooking and lighting. The Nepalese use water power mainly to drive mills for grinding grain, but it would be fairly simple to install a generator to provide electricity for lights and perhaps for cooking.

Two thirds of Nepal's people live in small hill communities so the idea of a national electricity supply is out of the question. The communities are scattered and small, the distances too large, and the routes too precipitous. Streams are the natural energy networks of the country. If small water mills could be sited close to villages, they could provide power for grinding grain, husking rice, and extracting oil from seeds over the entire country – without transmission lines.

The large diesel-powered mills that once served big communities have been gradually supplanted by smaller ones close to the farmers who need them. New machines that extract energy from a moderate water pressure, called cross-flow turbines, make it possible not only to replace the diesel engines, but to take the mills even closer to the users. Hardly any mills are still powered by diesel in Nepal.

Two factors bring the mills close to the consumer. First, fuel, by its very nature, is provided at the point of consumption. The mills are not dependent on long and tenuous lines of supply. One village, Majuwa near the town of Tansen, used to have a diesel. It took porters 2½ days to carry the fuel from Tansen. And each porter could carry only enough to run the mill for 5 days. In addition, turbine-driven mills are reliable and efficient. Streams that cannot easily drive a water wheel will drive a cross-flow turbine with only moderate changes to the stream's topology – for instance the addition of a small dyke. The remoteness of some mills was brought home to me by one of the engineers responsible for the manufacture and installation of small turbines. At the factory where they are made every machine is run in and checked on a test bed. "We are less worried about efficiency than by reliability," the engineer said.

The mills are the take-off point for electrification. Although many people assume that hydropower means electricity, in many cases – both in Nepal and other developing countries – a hydroelectric power station, no matter how small, is rarely what is needed most; nor can it be installed cheaply enough. Certainly in Nepal, villages have a greater need for mechanical power to improve milling, husking rice, and extracting oil from seeds. When a village has a mill for one or all of these processes it can then consider whether electric light is the next priority.

Lionel Makay, a water power expert working in Nepal, believes that electric lighting in itself is less beneficial than planners in many developing countries assume. It is difficult to quantify its convenience, of course, and it smacks of arrogance for people in better-off countries to question whether the Third World needs it. But without the improvement to the village economy that a mill can bring, electric lighting itself cannot materially enrich its users. But as soon as there is, say, milling to be done, or soap or paper-making industries are started, lights enable people to extend or arrange their working day with greater flexibility.

Different villages use different fuels for lighting. One tiny community burns freshly cut wood chips soaked in resin. This is an extremely profligate use of wood – but it costs no money. In another village, where the inhabitants use oil of mustard seed for lighting, a combined mill and generator could provide immediate and substantial benefits. Mustard oil is even more expensive than paraffin – but it is easier to obtain as the villagers grow it themselves. A water mill would provide them with a far more efficient seed press than the manual one they now use, and an associated generator would also give them electricity for lighting. The villagers could then produce more oil from the same plots, use none of it for lighting, and therefore provide themselves with a cash income, or at least a trading base.

But the biggest contribution that electrification could make to reducing the loss of forests would be in cutting down the amount of wood burned for cooking. Obviously the small generators that provide enough power for lighting would not be able to run cooking stoves. But cooking power suffers from the same problem as power for lighting; it is used only intermittently.

A possible solution would be to install a storage cooker that draws power when the system is not heavily loaded, and provide heat for cooking when necessary. The intention is the same as the storage heaters we have in Britain – they take electricity during the night when demand is low, and release heat during the day. Developing and Consulting Services, an organisation run by the United Mission to Nepal, is developing a cooking stove that might meet this requirement. Curiously, it is based on, and incorporates some parts of, a storage cooker found on a rubbish dump in Norway.

The Norwegians draw almost all their electricity from hydro sources, and at one time they too were experimenting with storage cookers to make better use of fluctuating demand. One way to encourage this is a tariff structure that encourages people to use small amounts of electricity continuously rather than large amounts at peak periods. Under the Norwegian system, householders could opt for a "maximum demand tariff". They would decide, with professional help, the maximum number of watts they would need to draw from the supply. Each household's electricity bill would be based on this figure only, and the consumer could use as much electricity as he wanted within the preset limit without paying more. An automatic cut-off for every house kept consumers inside the limit. To take advantage of this system a householder would want to adjust his consumption to use a small amount of power round the clock. This encourages the use of storage cookers, and the Nepalese authorities are looking to see if they could use a similar tariff structure when their own power supply takes shape.

Small water-powered generators, providing up to 5 kilowatts, could supply the Nepalese with power for both lighting and cooking if the storage device works satisfactorily. Low-voltage systems have the advantage that fairly unskilled people can install the equipment without danger. But they have the drawback that transmission losses are too high to supply houses more than 20–50 m away from the mill. The Nepalese will probably install generators that provide between 5 and 100 kilowatts at a conventional domestic voltage, using turbines small enough to need only simple civil works. Some installations already working incorporate simple dams that are swept away — by design! — with every monsoon, and are quickly rebuilt afterwards.

Water power has two other important uses apart from generating electricity. One is a more-or-less-developed technique, not commonly exploited, called "lift" irrigation. This uses energy extracted from running water by a number of devices — a hydraulic ram is one — to carry some of the water to a higher level. Water power can also generate heat directly through mechanical energy. Heat produced like this would be an important way of reducing wood consumption, particularly if it were used for drying crops. For instance, drying 1 kg of ginger needs 10 kg of wood. Development and Consulting Services is about to test a heat generator that is simply a very inefficient fan. The fan's blades force air through baffles, which heat up in the process, and the air emerges at a temperature of 150°C. The device could conserve a lot of wood burned on tobacco plantations — one Nepalese tobacco factory has 800 curing houses that are fired eight times a year, and burn 3½ tonnes of wood in each firing. Paper making, a local industry that could be expanded, needs heat for boiling water. Hot air blowers could at least reduce the wood needed for bringing the water to boiling point.

The Nepalese are not overlooking the role that biogas generators can play in reducing wood consumption. In Nepal, biogas generators are almost exclusively fuelled with cow dung (*gobar* in Nepali) and a small company indirectly supported by the United Mission is particularly active in selling and installing the so-called "*gobar* gas" generators. Many developing countries already use biogas. Whatever the country, the mechanics are pretty much the same. The fuel, almost any animal or human excrement, is mixed with water and allowed to ferment in a sealed pit or drum. The fermentation process gives off methane gas which is easy to collect or pipe to a cooker or lamp. The material remaining can be pushed out either by adding fresh dung or by gas pressure. Farmers can then use it as fertiliser.

In Nepal the dung comes from cows, which are sacred so there is no distaste in using the waste. Fermentation in the generator makes it as good, and sometimes better, a fertiliser than cow dung spread in the traditional manner. The biogas generators now installed in Nepal are certainly effective as cheap sources of cooking gas. But the Nepalese still have to solve certain technical problems. It is worth mentioning two examples because they illustrate very neatly the way in which technology has to be carefully combined with an understanding of local conditions.

The easiest way of using a *gobar* gas ring is to put the cooking pot on the grill before lighting the gas. You then adjust the air-to-gas mixture. But if you light the gas before putting the pot on the stove, the flame will almost certainly travel up the gas too quickly and extinguish itself. However, in Nepal, the rituals associated with cooking demand that the fire is lit before any pots and pans go on it. So engineers had to find a way of making a simple air/gas control that needs just the flick of a lever. This allows the cook to light the gas before putting the pot on the fire, and ensures that the gas stays alight.

The biogas engineers then faced the problem that the Nepalese cook with both round- and flat-bottomed cooking utensils. A cooking flame works most efficiently when the distance between it and the bottom of the pan is constant. The engineers adapted cookers to deal with both types of pan by developing a metal grid to fit over the gas ring. The grid has flat fins on one side for flat-bottomed pots, and inwardly-sloping fins on the other to heat a round-bottomed pot evenly.

Poor sales techniques can cause even good technology to fail. Most farmers, not only in the Third World, are suspicious of new technology, especially if it means investing money now against a promise of benefits in the future. Therefore when the first biogas generators are installed they must work properly not only for the owners' satisfaction but also to encourage others. Yet in some Third World countries the programme to encourage farmers to use biogas has not succeeded because the plants have failed for lack of a simple repair or adjustment. When this happens the owner becomes disillusioned and deters other people from turning to biogas. And he will often fail to pay back the money he borrowed to buy the plant. Yet a well-thought-out service arrangement can overcome this difficulty.

The Gobar Gas Company encourages its customers to pay a small amount extra at the time of purchase – which in any case is part of a loan – to take out a 20-year guarantee. This has worked for the 200 or so plants already installed. The guarantee is in three parts: it offers free replacement of any part in the first year, plus three visits by a technician; a further 6 years' guarantee covers the gas tank and masonry work (many gas tanks are in pits that are lined and roofed with masonry); and an annual visit by a technician for the next 13 years.

While the plant could supply gas for lamps as well as for cookers, it is five times cheaper to drive a generator with the gas and use electric lights. Diesel engines can be fuelled with many different gases – the company has found that a mixture of 20 per cent diesel and 80 per cent *gobar* gas gives the most effective performance. Thus the gas plant can fuel engines which may drive irrigation pumps or electricity generators, or both. And every village that converts to the new appropriate technology safeguards at least a small part of Nepal's woodland.

17

Kenya's growing concern over energy

"THIS WEEK"
7 June, 1979

Kenyan authorities are seriously looking at a new energy strategy, based on the traditional use of fuelwood, to solve their energy shortage and prevent a drift to a kerosene-dependent economy. All of Kenya's kerosene must be imported and this move to imported fuel could be disastrous to Kenya's future economy. If implemented over the next 15 years, the new strategy could reduce foreign currency problems, create up to half a million new jobs and possibly lower the price of charcoal, Kenya's main solid fuel, by a third.

The scheme was proposed at a recent workshop in Nairobi, attended by an international group of nearly a hundred top energy and environment specialists.

The workshop, on the implications of energy policies and their effects on the environment in East Africa, was sponsored by the Beijer Institute (Royal Swedish Academy of Sciences), the Kenya Academy of Sciences, and the UN Environment Programme.

After looking at the prospects for substituting hydropower and electrification for traditional sources of fuel such as wood and charcoal, as well as the possibilities for geothermal energy, alcohol, biogas, solar energy and coal, the group concluded that none of these, with the possible exception of coal, would be developed sufficiently to be a substitute for fuelwood within the next 20 years. And a new source of energy is urgently needed in Kenya within a decade or so. The proposed energy strategy could be applied across East Africa, for example, in Ethiopia, Uganda and Malawi.

About 70 per cent of Kenya's energy comes from fuelwood and charcoal – in some countries these fuels account for more than 90 per cent of energy consumption – but the existing rate of woodland depletion casts grave doubts on future supplies of fuelwood. Woodland removal in most parts of East Africa is already above "sustainable yield levels", a situation exacerbated by fuelwood lands being tilled for food as the population increases. Recent data presented at the workshop showed that for Kenya, by the mid-1990s, fuelwood will be confined to the low-rainfall rangelands. There the production of wood per unit area is so low, and the distances from the markets so great that many Kenyans will no longer be able to buy wood for fuel.

With the prospect of Kenya's population doubling by the year 2000, and the urban population growing by a factor of about six, the problem of how to find foreign currency to pay for the increased requirement of imported oil, with a global market already straitened by severe oil shortages, will become acute. Imported oil already accounts for about a quarter of the total Kenyan import bill each year. And the Kenyan fuelwood shortage will aggravate the world oil-supply problem. Already East Africans are substituting kerosene for wood.

The core of the new energy strategy rests not on new technology that is difficult for the farmers to understand, but on planting between 3½ and 5 million extra hectares of forests by the year 2000. The idea is to plant trees, not solely in huge tracts, as has been usual up to now, but also in small woodlots, adjoining a farmer's home or field. Such a rural afforestation scheme will involve community development officers, farmers, nurseries and the existing rural afforestation services. Another important new body, based in Nairobi, and formed under the impetus of Canada's International Development Research Centre, is the International Council for Research on Agroforestry. It has already played a strong role in the development of this new energy strategy.

Sadly, only a third of Kenya's land can be cultivated and, given the competition with land for growing food, there are doubts about the possibility of turning land over to forestry. Accordingly, research into the other energy alternatives will not be abandoned.

The workshop concluded that such a scheme would cost between £6 and £8 million each year, which may seem expensive. On the other hand, a kerosene-dependent economy would spend as much as four times this amount every year, drawn mainly from Kenya's scarce foreign currency reserves.

18

Kenya takes to the woods and the weeds

"TECHNOLOGY"
3 September, 1981

Kenya is considering two proposals to obtain energy from the land. In the schemes, forests could produce fuel for industry and a common latex-bearing weed could be cultivated to give hydrocarbons or small bricks of charcoal.

Shell, using data prepared by the World Bank, has proposed to Kenya's Ministry of Natural Resources and Environment that it should lease about 9000 hectares of an area that includes mature forest and recently planted eucalyptus trees. The trees would be processed into pelletised wood fuel that could be burnt in conventional industrial boilers.

After felling, the trees would be trucked to a nearby factory. This plant, largely based on paper-making techniques, would pulp the timber. The pulp would be Sun-dried and extruded through dies to form pellets of dried wood, about 2 cm long and 1 cm in diameter. These, says Shell, will provide the same energy efficiency as coal or oil and should cost slightly less.

After felling, the ground would be prepared and planted with fast-growing trees such as eucalyptus. Shell says that World Bank data indicate that the scale of felling and replanting would not significantly affect the water catchment of the area. Tea plantations, and other industries in the proposed forest, about 150 km north of Nairobi, would not suffer as a result.

The £4.5 million plan would save some £2 million per year in fuel imports. It would give 100 000 tonnes of pellets per year from 250 000 tonnes of wood.

On the negative side, however, the initiative seems to ignore the plight of poor people in Kenya who are short of firewood. And some ecologists think that industrial felling of mature forests could affect the watersheds and destroy unique ecosystems for ever. But Philip Leakey, a junior minister in the Ministry of Natural Resources, argues that the high cost of oil will hinder Kenya's development unless alternatives are found. He also says that fuel for industry has a parallel priority with firewood. The ecological questions remain unanswered awaiting a feasibility study that will follow if the Kenyan government approves the project.

Fewer pitfalls await a project that Leakey started himself. A fast-growing, robust weed called euphorbia could produce electricity and small lumps of charcoal, or a complex hydrocarbon that could be used to make liquid fuels or as a chemical feedstock. In conjunction with the Belgian government, Kenya has spent £250 000 on a year-long project to select the best natural forms of euphorbia, which Kenyans have used for years as hedges, and has done pilot experiments on turning the weed into hydrocarbons and charcoal briquettes.

The government has selected a type of weed that yields 80 tonnes of dried material per hectare, and thrives in semi-arid and arid lands. Each hectare of weed would produce each year about 20–30 tonnes of charcoal or up to 80 tonnes of the hydrocarbon.

The next stage is to commission and build a pilot plant capable of producing 20 000 tonnes of charcol, or the equivalent in hydrocarbons, per year and to continue with plant breeding to improve the yields of both products. Leakey believes that the venture is unique and that if it works it will bring three main benefits. First, it would let Kenya produce its own energy, a commodity of which it is woefully short; secondly, the scheme would generate a technology that the country can sell to other developing countries. Finally, the weed would restrict the downward spiral of arid land into desert. The euphorbia groups well at high density and prevents erosion of the soil.

19

Wood stoves: the trendy pollutant

MICHAEL ALLABY
and JIM LOVELOCK
13 November, 1980

Wood-burning stoves are fashionable, beautiful and friendly. They are also astonishingly dirty and Britain does not have enough timber to keep them going.

The wood-burning stove has come to symbolise our revolt against much that seems wrong with the modern world. It sits, attractive and comforting, a reminder of the possibility of better lifestyles, gracing our homes. It frees us from our dependence on technologies we can neither understand nor control, for we can gather its fuel for ourselves – at least in theory – and we can comprehend the growth of trees. It is economic, for wood is cheaper than its nearest rival, coal, and it conserves resources, for wood is renewable while coal is not. It is efficient, for it represents the sane, human application of modern design. It is socially and environmentally benign. Prettier than any electric fire its superiority in every other way is beyond dispute.

That is the myth with which we delight to deceive ourselves. The reality is sadly different. The efficiency of wood-burning stoves has been greatly exaggerated. In Britain, and even in the United States, timber resources are insufficient to permit the use of wood as a fuel except on a small scale. Finally, and perhaps most damning of all, the wood-burning stove is one of the most highly and most dangerously polluting domestic devices known to modern man. Not only is the banal electric fire its superior, it is its superior in every respect.

The myth is persuasive all the same. The stoves are often very attractive, their cast iron embellished with moulded decorations or vitreous enamel finishes. With their "traditional" appearance they conjure pictures of sweet-smelling smoke curling lazily from the chimneys of cottages tucked snugly in wooded valleys beneath towering mountains. They are artifacts taken directly from our fairy tales and folk legends and they appeal to our nostalgia. Aided by a little gentle marketing they have sold their way into more than 100 000 British homes in the past few years, while Americans have been buying more than 1.5 million of them a year.

They are not new, of course, but then they are not meant to be. Their design owes much to the Pennsylvania Fireplace invented by Benjamin Franklin in 1740 and no doubt, in America at least, their honourable ancestry enhances their reputation. Franklin's problem was to combat the cold of a Massachusetts winter and he needed something more effective than the open fire. The modern stove solves the same problem in the same way.

The inefficiency of the open fire arises from the flow of air that it causes and requires. As the hot gases rise through the chimney, convecting away almost all the energy yield of the fire, cooler air is drawn toward the fire at a lower level. The fire, then, needs a draught, and houses that depend on open fires are typically over-ventilated, with 10 to 20 air changes an hour in the rooms with fires. Although an open fire delivers radiant heat it does so only inter-mittently, when the fuel blazes or glows red.

The problem of over-ventilation is solved by sealing the fuel inside an airtight container, then controlling the flow of air by means of baffles. These can direct air through the burning fuel at a planned rate and in some designs they permit air to be preheated and mixed with unburned gases. This increases the efficiency of gas combustion, but sufficient unburned particulate matter is dis-charged to block the flues of most stoves. This accumulation of solids in the lower part of the chimney is characteristic of all wood-burning stoves and manufacturers are struggling hard to reduce it by improved design.

The result is that wood can be made to burn more slowly, so that stoves can remain alight for 12 hours without refuelling, while room draughts are reduced for the simple reason that the airflow through the combustion chamber is reduced. Even so, the effect is likely to be exaggerated. We are less tolerant of draughts than our grandparents were, and we try to reduce them. Modern houses are designed to eliminate unwanted draughts. Thus a modern house, or an old house that has been draught-proofed, will be more comfortable than an old house that has not be draught-proofed. The improved efficiency depends on the house as much as on the stove.

This is not a serious problem, of course, and in itself such a trivial

E

matter could be attributed to mild over-enthusiasm. It would not detract from the merits of the stoves. Their insatiable appetite for wood does.

The wood-burning stoves marketed in Britain are typically American or Scandinavian in origin, although they may be manufactured here. If we think for a moment about the countries with which we associate this technology – colonial North America, Scandinavia, Russia – we find it has been developed in regions with large resources of timber. Britain is not such a country and neither, any longer, is the US. With more intensive management than they receive at present, American biomass resources could supply some 10–20 per cent of total energy demand. If they are to provide more, then "energy farms" must be planted. On good land these will lead to all the problems we associate, rightly or wrongly, with modern intensively chemical- and energy-dependent agriculture. On very poor land stands of trees could help arrest erosion, but on marginal land intensive energy crop agriculture, based on annual species, might exacerbate it.

The US has exploited its timber resources to the limit, and today the only regions that have a reasonable hope of using fuelwood economically and sustainably are Canada, the USSR and tropical South America. The depletion of Asian timber resources, through commercial logging followed by slash-and-burn cultivation, is taking place on a scale large enough to be visible to orbiting satellites.

The difficulty arises from the relatively low energy density of wood. Britain's Forestry Commission points out that 1.25 tonnes of broadleaf coppice wood, dried to 20 per cent moisture and occupying 4 cubic metres of storage space, is equal to 1 tonne of coal, occupying 0.25 m^3. Broadleaved species produce denser wood than do coniferous species, but even among the broadleaves not all woods are equal. Poplar and willow give out less heat than oak and beech, for example, and stoves fuelled with them need stocking more frequently. Many of the brochures advertising wood-burning stoves recommend oak or beech.

The wood used for fuel must be dry. Damp wood generates larger amounts of tars to clog flues. Wood can be oven-dried, with questionable effects on the overall energy budget of the whole process, but in a well-made stack it takes about 12 months to bring green timber to a 20 per cent moisture content by air-drying. This is not necessarily a disadvantage if fuelwood is to be grown as a commercial crop, but for British enthusiasts it may present a new problem. Existing stoves are fuelled almost entirely by the large stocks of diseased elm, the product of Dutch elm disease, which sell cheaply because supply exceeds demand. We have no fuelwood plantations in this country. We could have them, but their produce would have to reflect production and transport costs. In the US, current growing costs run to around $25 per tonne of oven-dried wood, to which must be added a further $20–$30 for harvesting, transporting and processing even the most easily accessible timber.

Most of the advertising literature for stoves specifies the length of the logs they will accept but few of them mention the diameter. One that does suggests 13 cm diameter wood. This is sensible enough while the source is elm, but it would hardly be economic to grow wood to this thickness deliberately as a source of fuel.

The most sensible strategy devised by the Forestry Commission would be to plant stands of timber for coppicing. In this way 5 cm diameter wood can be harvested at about 4–6 tonnes per hectare per year from oak and beech. The faster growing willow and poplar yield 12 and 16 tonnes per hectare per year respectively and they can be harvested usefully after about 10 years compared with the 15–20 years it takes to establish maximum yielding stands of oak or beech; but because their fuel value is much lower they offer no real advantage. Yields can be increased by harvesting wood with a smaller diameter, but only at the cost of increasing handling difficulties and using more storage space.

It is on the basis of these figures, derived from experimental plantations, that the Forestry Commission has calculated the area of coppiced plantation needed to keep the wood burners supplied. A three-bedroomed house, heated entirely by wood-burning stoves, needs 3 hectares of woodland, of which 0.2 ha must be cut and processed each year. This sounds little enough compared with the 1.25 million ha of forest in Britain, but if we suppose that the existing 100 000 or so stoves require 300 000 ha of woodland, this is equal to the entire forest area of England. (Most of Britain's forest is in Scotland.) Obviously, Britain will find it difficult to produce fuelwood on this scale.

Despite its renewability wood, like all biomass fuels, can be produced on a large scale only by diverting land from other uses. For example, the land needed to produce material for alcohol to keep one motor car on the road might otherwise feed 8–16 people. Current "gasohol" production in the US is fed by withholding exportable grain and the world's largest producer of fuel alcohol, Brazil, is also the world's largest importer of grain.

Nevertheless, some people may argue that the price is worth paying if wood is even a partial alternative to, say, nuclear power because, after all, the burning of wood creates no serious environmental problem.

There is some truth in this, but very little. It is true, for example, that wood contains little sulphur, so its combustion will produce little sulphur dioxide. It is true, too, that it contains no radium, so there will be no emission of radon and its products – although there may be low levels of radioactivity in the ash. Because wood is re-grown, the carbon dioxide produced by combustion is balanced by carbon dioxide absorption in the fuelwood plantations, so wood burning does not imply possible climatic effects from the release of this gas. In this respect wood is superior to fossil hydrocarbons. There, the advantages end. Wood smoke is rich in polycyclic organic matter.

The list of products of wood combustion includes such compounds as benzo(*a*)pyrene, dibenz(*a,b*)anthracene, benzo(*b*)fluoranthene, benzo(*j*)fluoranthene, dibenzo(*a,l*)-pyrene, benz(*a*)anthracene, chrysene, benzo(*e*)pyrene and indeno(*1,2,3-cd*)pyrene and many others less noxious. The ones we have named are all known or suspected carcinogens and if the list sounds familiar, so it should: it is the list of substances found in tobacco smoke and suspected of causing lung cancer. This similarity should not surprise us, for one vegetable product is very much like another chemically, and in recent years there have been campaigns to discourage people from lighting garden bonfires because of the pollution they cause. A bonfire burns mainly wood.

Is it fair to compare a lovely, well-made modern stove with a smouldering heap of garden refuse? No, of course it isn't. On balance the stove is the more serious offender. This is because the emissions consist of unburned hydrocarbons. They are fuel that has not been used. The way to improve combustion efficiency is to increase combustion temperatures. The wood-burning stove is designed to burn its fuel slowly, and at low temperatures, and so its rate of emission increases inevitably. The more efficient the stove, the more pollutants it releases. Indeed, in extreme cases the stove that is damped down to make it stay barely alight while its proud owners go out for the day may behave like a wood gasifier, or pyroliser, heating the wood in the absence of air to produce greatly increased amounts of organic matter for discharge into the outside air.

Coal is rather similar to wood, and so we may expect coal fires to cause pollution problems similar to those caused by the burning of wood. In 1979, though, the publication of the results of a study commissioned by the US Environmental Protection Agency (EPA) and performed by Monsanto Research showed that there are substantial differences in the environmental impact of the two fuels, due not to their composition but to the way they are used. Where polycyclic organic matter emissions from wood were given a value of 1, those from residential coal fires scored between 0.33 and 10; but those from coal-fired utilities, industrial coal, and oil-fired boilers and furnaces scored between 0.01 and 0.33. In other words wood burning causes more serious pollution than anything except a dirty, smoky coal fire and, in the words of the report, "carbon monoxide and POM emission rates, expressed as grams of pollutant per kilogram of wood burned, were an order of magnitude higher from the stoves than from the fireplace".

Again, there is nothing very new or startling in the discovery that fuel combustion on a small, decentralised scale causes more pollution than the combustion of similar fuel in a large, more efficient furnace. Industrial users of fuel can be required to install devices to clean waste gases and remove particles before discharge, and they are likely to be more concerned about the waste of costly fuel that inefficient combustion implies. Wood-burning stoves, on the other hand, are installed in private homes, in the most decentralised manner imaginable. That is their principal attraction. It would be quite unrealistic to suppose that domestic users could be required to fit flue scrubbers or electrostatic precipitators, even if such devices could be made to work in a small chimney. The issue was summed up in a paper produced for the EPA by the Battelle Institute: ". . . small, low-temperature wood-burning units tend to produce more undesirable atmospheric emissions than do the larger combustion units which operate at higher temperatures and with greater turbulence . . . it logically follows that any potential threat that may result from wood combustion will come from the residential, not the industrial sector".

In Britain wood-burning stoves are still few and far between. The dangers are potential rather than real. In the US, though, their use may have to be controlled in New Hampshire, Vermont and parts of Colorado. In some New Hampshire valleys, especially prone to haze, atmospheric particulate matter from wood burning has reached concentrations high enough to place a restriction on industrial expansion. The EPA is loathe even to contemplate any general regulation of a purely domestic activity, but the problem of wood-burning stoves may well become acute. Their environmental effects may limit their use.

It is inconceivable that wood burning in Britain will ever cause pollution problems like those we used to associate with coal fires. Unlike the Americans, the British government legislated, and the Clean Air Act forbids the use of smoky fuels in specified areas that are particularly at risk.

However, it is not the law that will impose limits on the use of British stoves, but the availability of fuel. Within the next year or two, or whenever the stocks of elm are exhausted, owners of wood-burning stoves in this country will have to turn to new sources. Perhaps fuelwood plantations will be grown for their convenience, but even then it is doubtful whether they will be able to supply the stoves that are in use now, let alone any that may be bought in the near future. People will face a choice between very high fuel bills, probably for imported wood, or the need to find new uses for what suddenly transform themselves into pieces of decorative, but quite useless, ironmongery. It is not beyond the bounds of sinister possibility that wood stolen from forestry plantations might command a price high enough to sustain a black market among people who have invested hundreds of pounds they cannot recoup. Thus is innocence lost!

There is a place in the world for wood-burning stoves, but it is a small place. For people living in rural areas with access to waste timber in quantities sufficient to ensure a regular supply such stoves may make good sense. After all, they do provide space and water heating. They create problems only when they become too numerous. They can contribute nothing to the energy supply of the densely populated and sparsely forested industrial countries.

There is a sad irony in the discovery that so attractive a device, such as embellishment to any home, far from enconomising on the use of scarce resources while protecting the natural environment in fact squanders resources and causes environmental damage on a scale that would appal the manager of a factory or power station.

<h1 style="text-align:center">20</h1>

The wood-burners strike back
27 November, 1980

Allaby and Lovelock's article on the less pleasant aspects of wood-burning stoves evidently touched a raw nerve. This is some of the correspondence that resulted.

Wood-burning stoves

Relaxing in what has become the warmth and comfort of my own home since I bought a wood stove, the intrusion of your 13 November issue is not entirely welcome.

While I do not take issue with all the article on wood stoves, by renegades Michael Allaby and Jim Lovelock, my new-found friend has been rudely insulted and must be defended.

I have the good fortune to live in an area which grows wood better than anything else and also is thinly populated. While the awful destruction of the native forests is never far out of mind here, for the time being at least, forest is on the increase. In fact, I planted some 50 acres (20 hectares) myself this year. So far as I am concerned, wood is a renewable resource.

The woodstove is a replacement for an open fire and viewed from this perspective it is a miracle of ancient technology. For me it has turned a draughty, damp and cold house into a very comfortable one, at the same time reducing the time spent fetching and sawing wood, and saving the money spent on coal to keep the fire going. Apart from its miserly consumption of fuel, it is a very efficient convector and produces comfort in a house of a type which it would be hard to produce with gas or electricity.

Wood-burning brings home one of the truths of the energy crisis. When the cost of heating the home is measured in terms of physical effort and time, rather than money, the consumer can make a better decision on how much heat is needed. It is easier to economise when the cost is in effort and time which is immediate rather than in distant fuel bills. Wood-burning teaches us that the best way to conserve energy is to use less.

To supply an average centrally heated British home with fuel for cooking, hot water and heating requires a little over 70 GJ/year as gas or oil combusted with some 90 per cent efficiency. Wood at 20 per cent moisture yields about 13 GJ/tonne and, if one assumes that its combustion can only be 60 per cent efficient, about 8 tonnes/year of such wood is required for the average home. This quantity, in poplar or willow (using the mean of Allaby and Lovelock's range of yields, 12–16 tonnes/ha/year), can be produced annually from 0.57 ha, not 3.0 ha as the writers say. Hence, the 100 000 existing stove-owners would need 57 000 ha of forest to supply them, not 300 000 ha.

The total woodland area of the UK is over 2 million ha, not 1.25 as stated, so the impact of the existing stoves on the potential supply is really rather small – only 2.8 per cent at most. Because a great many wood stove installations do not extend to full central heating, the actual requirements must be well below those figures.

Finally, no account was taken in the article of the potential for extracting presently unused wood-waste from the forests that are producing timber for industrial uses. Analysis shows that some 250 000 tonnes/year would be fairly readily available now, growing towards 500 000 tonnes/year when the younger forests mature. We also have considerable scope in Britain for increased forestry area. Therefore, provided some coppice is planted soon for fuel purposes, there seems little harm yet in encouraging the sale of a few more hundreds of thousands of wood-burning stoves.

21

Why wood is not for burning

KEITH WILLIAMS

15 October, 1981

As Allaby and Lovelock pointed out in Chapter 19, wood stoves are not the benign technology they are often claimed to be. Part of the problem is that traditional forestry in Britain wastes something like 40 per cent of the biomass material that trees produce. Turning this waste into charcoal, artificial logs or even alcohol might increase the efficiency of growing wood for fuel.

Wood, of course, is one of the least suitable fuels for wood stoves. Paradoxically, this truth is emphasised by our annual wastage – or, to put it another way, refusal to consider the use – of several million tonnes of burnable wood every year. This is enough to heat about half a million homes.

Nobody would suggest that wood fuel is the only replacement for fossil fuels when they are exhausted. But in an energy-hungry world there is at least enough of it growing for us to discard our unrealistic concepts and come to terms with its real characteristics. Even those manufacturers and importers thriving in the ebullient market for wood stoves have little long-term confidence in their future. Many believe that, any year now, Britain will be running out of home-grown wood supplies. Whenever a local shortage occurs, due not to a shortage of raw material but to the lack of any way of distributing and marketing it, the pessimists mutely accept this as the beginning of the end.

Anti-wood activists publish blatantly propagandist pictures of bare tree stumps, lined up by the hectare as headless witnesses to the savagery of the feller's chain saw. They fail to mention that these are broadleaf species that will naturally regenerate by coppicing. The same site some 15 or 20 years later will show the woodland much restored with vigorous crowns growing from the boles or stumps. Conifers, too, grow back by reseeding at a rate that would surprise most of these people.

In Britain, we use about 45 million tonnes of forest products a year, but only 4 million tonnes comes from indigenous sources. The last woodland census, held in 1966–67, showed that Britain had 1.85 million hectares of trees. Since then, replanting has continued to increase the stock of timber by up to 40 000 ha per annum. An aerial survey done last summer – but not yet published – is expected to show that the "hectareage" is now over 2 million.

Among the slowest growing of British trees is the oak, which on average yields 2.4 tonnes of new wood per hectare each year. Conifers grow up to three times as quickly: for example Douglas fir produces about 8.3 tonnes of new wood every year and so does western hemlock. Some poplars register even higher rates of growth. As conifers constitute about 60 per cent of the nation's trees it seems reasonable to take a mean figure of 4.2 tonnes per hectare as the annual growth figure. On this basis the wood crop totals a minimum of 8.4 million tonnes each year. Deducting the 4 million which already go for commercial use, 4.4 million tonnes remain.

I have established by testing that you need 5–7 tonnes of wood to maintain a temperature of 18°C in a three-bedroom house all year round. Taking the higher figure the 4.4 million tonnes of wood that Britain fails to use could supply full household heat for some 630 000 homes. Allowing for every kind of shortfall in supply, my previous estimate of a half million homes fuelled with wood seems quite reasonable.

Of a quarter of a million wood stoves thought to be working in Britain, most are running on a "part-heat" basis and certainly consume no more than 4 tonnes of fuel each year, if that. Assuming this is taking up a million tonnes of the surplus, more than 3 million tonnes remain.

If any of it could reduce imports of commercial timber it would have been brought to market already. Mostly it is in remote woodlands too small to be profitably harvested, or consists of sizes below those that commercial firms find acceptable. Often the owners of small woods find that the prices being offered for firewood do not cover the cost of cutting and delivering it. But as prices rise, more of these supplies are beginning to be used.

It is at this point that one realises how inadequate a fuel raw wood is, and how unsuitable it is for burning in a modern wood stove.

First, it poses, on site, an intractable transport problem. Gathering loads of small wood is time-consuming and labour-intensive. The wood itself is bulky and awkwardly shaped. A lorry load of it includes a lot of air. Even with straight-limbed trunks the air space in a load is about 30 per cent of the total. With "rough tumble" loads the figure is more like 50 per cent. Once this wood supply is delivered to a household it must be stored and air-dried for a minimum of 6 months (if it is green, or sap-laden wood). A year's supply will fill a large domestic garage, or its equivalent – about 55 m³ of space.

Then you must design a suitable appliance to burn the wood. For the woodstove designer the characteristics of the fuel he must work with are a nightmare. Unlike coal – where the consistency, grade and size of the fuel can be predicted to a nicety – woodfuel remains an almost entirely unknown quality. It may be anything from green to rotten. Within that range its chemical constituents will vary widely. Green wood is half water, while the "overdry" woods that come from small woodworking factories contain less than 10 per cent of water. As wood dries and begins to rot, so the chemicals in it – acids, alcohol, carbonyls, hydrocarbons, phenols and other substances – will dissipate or even vanish entirely. The designer cannot even guess what size of wood his creation will be expected to burn, let alone what types of timber. Inevitably he must try to guess his way through a series of compromises and hope that his formula will satisfy some of the demands made of it.

Even worse, when he is trying to plot his burning calculations, he will find little detailed, reliable information describing the endothermic and exothermic sequences of pyrolysis. Nor is there any laboratory facility in which he can test the concepts incorporated in his design.

Firewood is not designated as a "smokeless" fuel within the definitions of the Smoke Control Acts, so that its use is prohibited in areas subject to such control. This is an unfortunate consequence of a bias built into the procedures used to classify different fuels. The criteria are designed to fit the characteristics of coal. Wood, when it is first lit, emits quantities of a pale grey gas classified as "smoke". In fact much of this is no more than water vapour driven from the wood fibre by heat or generated as a chemical by-product of the burning process. The exact proportion of water vapour remains a subject of debate and testing, as it varies with the constituents of the wood. However, when the wood is at about 500°C, and the pyrolysis cycle (literally the cycle of breakdown by heat) is complete, the fuel being burnt is virtually charcoal; in other words it is about 75 per cent pure carbon and classifiable as a smokeless fuel. This occurs about 15 minutes after a wood fire has been lit.

Between the competing claims of the manufacturers for the performance of their stoves (which are hard to compare because different makers quote different criteria), and the variety of advice they give as to how to operate them, are warped by the retailers who sell the stoves, and the consumers who buy them. Few in either of these groups have any previous knowledge or expertise in the management of wood fires. As a result many stoves work at well below their potential best performance.

These points summarise the main arguments against wood as a fuel. Yet there remains a strong incentive to solve them if these surpluses of wood fuel exist in the abundance that I claim. Can this be demonstrated by leg work?

I have tramped through every kind of woodland from Devon to Aberdeenshire. The dominant impression is of the profusion of its supply, and the wantonness of its waste. A variety of wood waste lies on the ground – anything from twigs and small branches up to main stems and collapsed trunks. Dead limbs wait to be trimmed from trunks and dead trees whose roots have long since rotted away lean drunkenly for support against ther neighbours. Disease and wind damage will produce anything up to ½ tonne per hectare of such standing dead wood in scrub woodlands every year.

But you find the most needless man-made waste in managed woodlands; in other words it is the result of deliberate policy. The Forestry Commission is not free from blame in this respect, although its rationale makes some sense. When a tree is felled – or a stand of trees is felled – only the main trunk and branches are reckoned to have commercial value. The remaining small wood – the so-called lop and top – is trimmed off. It cannot be left to lie and rot as this encourages weeds (although I have often seen such trimmings that have been left lying for more than 3 years). The usual remedy, adopted for example by the Forestry Commission, is that the lop and top is burnt by the fellers, who charge higher piece-work rates for doing this work. Alternatively, the owner of the wood will pay someone to dispose of the lop and top at a fixed rate per hectare.

The lop and top amounts to about 40 per cent of the total bulk of the tree. And it is this and other sources I have mentioned that make up a large part of the 4.4 million tonnes of annual growth that is wasted. Yet, how are these supplies to be brought to workable firing? To gather them by hand would result in a market price higher than that of solid fuel – no consumer would be willing to pay it.

Some way must be found to convert these waste woods into saleable fuels.

At present there are three main possibilities, although two are probably interim solutions. Both in Europe and North America (but never until now in Britain) companies have had some success in making artificial wood logs. The American Presto-log and the Swiss Glomera processes are similar. First, the firms reduce the wood to chips and then subject it to high pressures (about 20 685 kN/m²) in a mould through which it is moved at a steady rate. Putting wood under this pressure raises it to a high temperature and the natural resins in the wood plasticise. When the pressure is suddenly released the resins solidify and hold the wood to the shape of the mould. The process produces continuous tubes of wood which can be cut into standard logs, Unfortunately, commercial firms have never accepted this technique as other than a cosmetic fuel and they sell the "logs" at prices higher than for solid fuel.

The second of these intermediate solutions centres on converting wood to charcoal. One way of surmounting the handling costs on small wood is to convert it to charcoal on-site in portable kilns. One worker can produce up to 5 tonnes of charcoal from each kiln per week, and can work more than one kiln at a time. At present, firms in Britain import charcoal of an unreasonably high specification and sell it in small sacks at about £750 per tonne – in other words at £2.25 for a 3 kg bag. My own calculations show that you can make a lower grade charcoal from mixed small-wood supplies and sell it at about 70 per cent of the price of coal. Alternatively, you could pulverise this charcoal on site, and transport it as a dust to central depots where you could, by adding a binder such as starch, convert it into briquettes.

However, there is a third possibility with more long-term potential in which R&D activity proliferates. Both Canada and the US are now self-sufficient in energy supplies, with one exception – fuel for transport. Brazil's economy hinges on whether an alternative to liquid petroleum is available. The Brazilian government believes that fuel-alcohol should fill that gap; its engineers and scientists are now the world leaders in producing fuel-alcohol. Brazil now has over 200 fuel-alcohol plants in production, and another 190 are planned. Last year Brazilian plants distilled about 5000 million litres of ethanol from sugar cane and cassava. For several years all new cars in Brazil have been designed to burn the 20 per cent gasohol mixture – one part of ethanol mixed with four parts of petrol. Now, the Brazilian government is stimulating a strong market demand for cars that will run on straight alcohol of 192 proof.

The US is treading hard on Brazil's heels. Its Department of Energy calculates that alcohol production reached 500 million litres in 1980, and predicts that this will rise to about 4500 million litres in 1982. Literally dozens of new distilleries are being built or are planned in the US specifically to produce fuel-grade ethanol. They use corn, barley, potatoes and other vegetables as the raw materials.

Canadians have a certain distaste for using food stuffs as a source of liquid fuel and plan instead to utilise available forestry resources, such as unused sawmill wastes, small wood remaining after tree harvesting, non-commercial tree species and forests in unworked regions. Canada's energy department believes these sources could yield 121 million tonnes of dry wood each year, and that could be processed into 50–70 million tonnes of fuel alcohol. Fuel grades lower than those required for internal combustion engines would be suitable for central heating boilers in homes, factories and offices.

Clearly the modest 4.4 million tonnes of waste wood available in Britain will not go far toward converting our total energy requirements, even with the additional million hectares of trees the government intends to plant. But a diversity of fuel sources that includes trees might help to avoid the inflationary monopoly that has scarred these past years of the liquid fossil fuel era.

PART FIVE

Charcoal makes a comeback

The future of charcoal as a fuel is of course tied to that of its raw material, wood. But unlike wood, which is usually a rural fuel, the bulk of charcoal is consumed in towns and cities. Its advantage, of course is that it is lighter and more compact than raw wood, so it is a better proposition to transport. Therefore it is the simplest, and most common, commercial fuel derived from processing biomass.

Unfortunately, turning wood into charcoal can be a very inefficient way of tapping its energy. The most widely used method of making charcoal is to stack wood into piles, cover it in earth, and let it smoulder for a few days. The process wastes at least three-quarters of the wood's energy. Steel kilns, such as the ones being introduced into Sri Lanka (p. 125) are far more efficient, but are expensive.

One way to overcome the apparent disadvantages of making charcoal is to use agricultural or forestry wastes as the starting point. We described one possibility in "Why wood is not for burning" (Chapter 21). Another possibility would be to make charcoal from the waste of industries such as sugar cane.

The third article in this Part illustrates the potential value of charcoal as the feedstock for gasifiers to drive vehicles. Such gasifiers were common sights during the Second World War, but their unreliability drove them off the streets when cheap oil became available. However they could be set for a revival in some parts of the world.

22

Charcoal out of sugar cane

"SCIENCE IN INDUSTRY"
30 May, 1957

Bagasse, the crushed sugar cane after the sugar has been extracted, can now be converted into high-quality charcoal, according to Mexico's Department of Agriculture. The discovery is important to Mexico because all efforts to check deforestation by legislation and by encouraging the use of oil for cooking have had little success.

The researchers report that charcoal obtained from bagasse burns easily, makes no more smoke than wood charcoal does, has the same heating value and gives off a "light, pleasant odour". It can be moulded and compressed to adjust the density and so control the rate of combustion. An industrial process for "practically automatic production" has been worked out and various grades of this charcoal are to be put on the market.

In their search for an alternative to wood charcoal, the government researchers experimented with chips, sawdust, hay, corn husks, the blades of wheat and rice, and other agricultural residues. Bagasse proved the most practical of all.

23

Sri Lanka's energy problem comes out of the woods

"TECHNOLOGY"
13 August, 1981

Sri Lanka could soon become an energy exporter with a scheme that will also increase its area of arable farmland. The country's Foreign Investment Advisory Committee plans to turn an embarrassment of waste wood into charcoal. The fuel could power the local cement industry or could even be exported, for instance to Japan and Saudi Arabia.

During the past few years, Sri Lanka's foresters have cleared the giant Mahaveli forest to create new, irrigated croplands. Although cash for the project has been no problem – the World Bank has helped out – the Sri Lankans have been left with a lot of wood from felled wira trees. Wood, together with kerosene, is the nation's main domestic fuel, but the 25 million tonnes of wood that the forest will provide within the next 5 years is a long way from towns and villages, and so is too expensive to transport. Traditionally, foresters burn it on the spot or leave it to rot.

But when it is converted into charcoal, the wira becomes a much lighter and more efficient fuel. So a new company, Charlanka, will work with Enterprise Development Incorporated of Washington DC and the Sri Lanka State Timber Corporation to exploit this rich source of energy.

To help the Sri Lankans to develop an efficient means of producing charcoal, Garry Whitby of Intermediate Technology Industrial Services (ITIS) in Rugby has supervised trials with wood kilns this year. ITIS collaborated with Shirley Aldred Engineering, a British company that makes a portable steel kiln, in cooperation with Peter Hock, a consultant. "The new kiln is expected to produce around 800 kg a 'burn', although tests have given 900 kg of usable charcoal," says Whitby.

Each kiln is an inverted cone of steel rings. It lasts for 3 years, and costs about £500. Villagers would find them more efficient than traditional earth furnaces. Charlanka also believes that the cost of making charcoal in the kilns will encourage the Sri Lanka Cement Corporation to substitute the fuel for some of the 300 000 tonnes of coal it imports each year. And the company also estimates that the kilns will produce a surplus of 15 000 tonnes of charcoal each year for export.

The timber corporation has joined the scheme to promote sales of its own wood-burning stoves. The government recently removed all subsidies from sales of Sri Lanka's other important domestic fuel, kerosene. But the country's Institute of Scientific and Industrial Research cites important medical reasons for promoting charcoal as a smokeless fuel: it could, claims the institute, cut the number of cases of eye, skin and lung diseases that Sri Lankan women commonly suffer in unventilated kitchens.

People who live within 200 km of the kilns will receive the charcoal; the costs for transporting wood become too high when forests are more than 80 km from a village. This will make the Mahaveli's charcoal available to 200 000 people and will provide more employment in rural forest areas, although Charlanka has not yet decided who will own and run the kilns.

Not all of the cleared forest will become farmland. Within 5 years, foresters will plant 14 000 hectares of fast-growing eucalyptus trees to replace the wira as sources of charcoal run out.

24

Big revivals for biomass engines

GERRY FOLEY and GEOFF BARNARD
17 February, 1983

Independent and major revivals of Second World War gasifier technology are under way on opposite sides of the world. A gasifier is a
furnance which is bolted to a vehicle, and produces a combustible
gas by burning biomass in a restricted air supply; the gas replaces
petrol or diesel. The technology kept basic transport running in
Europe during the 1940s, but it is now being redeveloped in distinct
ways by the Philippines and Brazil.

The Brazilians rely on private enterprise. Some 60 companies
there claim competence and expertise in the technology, each trying
to identify its own market niche. So far about 10 have made sales.
Industria Siquieroli has made and sold the most units – 650 charcoal-
burning vehicle gasifiers at $750 a time.

Other firms specialise in direct-heat gasifiers to make gas to fire
furnaces and kilns. GTI has particular expertise in wood-burning
gasifiers for making lime: its largest installation consumes about
0.8 tonnes of wood per tonne of lime produced; traditional
methods consume up to 2.5 tonnes of logs for the same result. The
gasifiers provide up to 6 million kcal/hour.

The Philippine government financed the Gasifier Equipment
Manufacturing Corporation (GEMCOR) in 1981. It sold 850 units
in its first year. Of these, more than 450 were for small, inshore
fishing boats called bancas. Nearly 200 were built to power irrigation pumps, and most of the rest were for light commercial vehicles.
GEMCOR now works a three-shift, 24-hour day and employs 125
workers. Production should reach 4000 gasifiers this year, and
orders are flowing in.

Gasifiers disappeared in Europe after the war when oil became
available again. They were not an attractive, efficient or safe way to
power a vehicle. But the oil price rises of the 1970s created conditions in many Third World countries similar to those of a wartime

siege economy. The world may be awash with oil, but some countries pay half or more of their export earnings to import it. In rural areas, irrigation pumps and tractors stop because farmers cannot afford fuel.

Despite the problems revealed by wartime experience, gasifiers have tantalised researchers since the early 1970s. They offer a way to substitute local wood and charcoal for expensive imported diesel and petrol. More than 100 development groups throughout the world are working on the technology, many of them attempting to update old designs to work with modern engines. Several of the gasifier systems commercially available are made by Second World War manufacturers. They have dusted off their old blueprints – or even their old stock – and entered the market again.

Sponsorship by various technical-help agencies has allowed prototype gasifiers to be installed in Tanzania, Guyana, Tahiti, Seychelles, Indonesia and elsewhere. Representatives of the EEC's technical assistance programme have been promoting the technology in Pacific islands such as Fiji and Tonga. But, despite such efforts, the technology has not taken off commercially. The best installations have demonstrated what is already known – that gasifiers can be made to work; the worst have been total failures.

The reasons for such poor results are simple. Gasifiers need considerable technical skills for operation and maintenance. A poorly-run gasifier rapidly becomes unusable and may destroy the engine to which it is attached. The technology, simple in principle, is not a typical Third World village technology because, for successful deployment, it requires technically sophisticated operators as well as repair and maintenance technicians.

A gasifier is a furnace in which the supply of air is restricted to prevent complete combustion, hence its more descriptive tag "gasification by partial combustion". In principle, any reasonably dry biomass can be gasified; but the practical problems of using fuels with a high ash content or of low density are quite severe. Charcoal, wood chips, coconut shells and maize cobs are the only practical fuels at the moment.

Typical commercial gasifiers consist of a reactor, where the gas is produced; a gas cleaning and cooling system; and a modified carburettor to feed gas and air into the engine. In most, the suction of the engine draws air into the gasifier and pulls the resulting gases through to the piston chambers. The typical "downdraught gasifier" (there are also updraught and crossdraught models, see diagram) has its combustion zone, or "fireball", in its lower part. Air is drawn

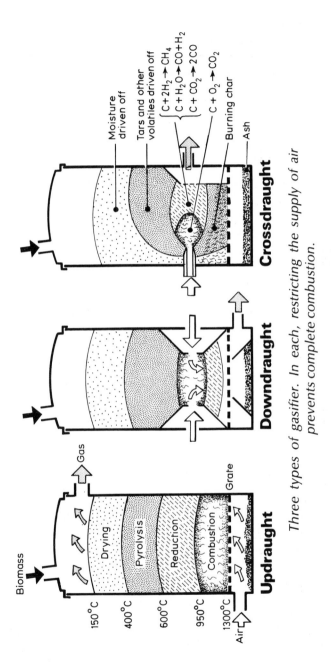

Three types of gasifier. In each, restricting the supply of air prevents complete combustion.

Updraught

Biomass

Gas

Drying — 150°C
Pyrolysis — 400°C
Reduction — 600°C
Combustion — 950°C
Grate — 1300°C

Air

Downdraught

Crossdraught

Moisture driven off

Tars and other volatiles driven off

$C + 2H_2 \rightarrow CH_4$
$C + H_2O \rightarrow CO + H_2$
$C + CO_2 \rightarrow 2CO$

$C + O_2 \rightarrow CO_2$

Burning char

Ash

in and the wood or charcoal burns, producing water and carbon dioxide at temperatures of 1100–1700°C.

The supply of air is restricted so the combustion zone can spread only so far. The hot gases produced are sucked downwards, out of this zone, into a slightly cooler area which contains unreacted carbon from the feed material. No free oxygen is present so a different set of reactions takes place; the atoms rearrange themselves, seeking the most thermodynamically stable combinations. Most important to the user, carbon dioxide is converted to carbon monoxide and water breaks down to give hydrogen. The oxygen atoms released combine with unreacted carbon to give more carbon monoxide. The final gas mixture, called producer gas, contains around 20–30 per cent carbon monoxide, 5–15 per cent hydrogen, along with traces of methane. The remainder is largely nitrogen, with some carbon dioxide. Because nitrogen is inert, the calorific value of the mixture is low (3000–5000 kJ/m³) about one-tenth that of natural gas. The mixture's high carbon monoxide content makes it lethal. Inhaling it can cause instant death; the effect has been described as comparable in speed with an acute cerebral haemorrhage or coronary thrombosis.

The combustion zone heats the unreacted feedstock that surrounds it. Where the temperature exceeds about 400°C the feedstock begins to break down, giving off water vapour, methanol, acetic acid and – of most concern to the mechanic trying to keep the contraption going – a lot of heavy hydrocarbon tars. The amount of tar may be as much as 40 per cent of the weight of the original feedstock. Charcoal has these tars driven off in the course of manufacture, leaving it as a clean, uncomplicated fuel for gasifiers.

With other organic fuels the problem is to reduce the quantity of tar in the final gas to a level that the cleaning system can deal with. Most gasifiers are therefore designed to draw the tars through the combustion zone so that they are cracked or burned. Excessive moisture, the formation of clinkers, big load-changes, and a variety of other circumstances, however, can let some tar through.

After leaving the gasifier, the gas is cleaned and cooled. A cyclone is generally used to extract dust and ash. This is followed by a series of filters and scrubbers. Cooling is usually done by passing the gas through pipes. In the case of a vehicle, these may be at the front of the radiator or on top of the cab to take maximum advantage of the airflow.

The cleaned and cooled gas can power diesel or petrol engines. In the case of compression-ignited diesel engines, a small amount of diesel fuel must be injected in each firing cycle to allow ignition. Thus the maximum saving of diesel fuel is about 85 per cent. In practice the saving may be considerably less, because extra diesel is usually needed to take up any temporary increases in the load on the engine.

Gas can substitute entirely for petrol, because ignition is caused by a spark-plug. In practice, however, most engines retain a means of switching to petrol for starting or to meet heavy-load surges, so the fuel saving is less than 100 per cent. Converted petrol engines suffer a considerable loss, around 50 per cent, of maximum power.

Running a gasifier requires both patience and mechanical aptitude. In starting, a fire must be kindled and the gasifier nursed up to full output. Filters must be cleaned, cooling systems drained, and the whole assembly checked regularly and carefully for gas leaks. Above all, operators must be able to adjust and tinker with air, gas and fuel flows to keep the systems running happily. Gasifiers yield nothing but trouble if they do not get continual care and attention. This is not to put gasifiers in a realm of esoteric complexity. They require the care that can be provided by a loving diesel mechanic; unfortunately, such people are rare in developing countries.

Reviewing the progress in promoting gasifiers during recent years, it is easy to despair and conclude that, as with many other energy initiatives, the practical difficulties in mounting an effective programme are virtually insurmountable. Yet clearly the Philippines has made a breakthrough. Much of the credit must go to Ibarra Cruz of the University of the Philippines and the head of the national non-conventional energy centre. Cruz built his first Philippines gasifier in 1967, and through the 1970s his research has resulted in a series of models of increasing practicality, simplicity and cheapness. These are now the basis of GEMCORs range.

Three basic types are produced. The smallest, to power a 12 kW engine for a fishing boat, is light and compact. It is designed to take minimum space and weighs 55 kg. A wheelbarrow-mounted version can be switched between different small engines in a farm or industrial plant. Irrigation pump models are for engines of up to 45 kW. These have to operate for prolonged periods and are robust and solid pieces of equipment, for space and weight are not serious constraints. The fact that water is available is used to advantage in a simple water-scrubber for cleaning the gas.

A lightweight, modular unit has been designed for vehicles. It can be fitted to a jeep or truck without extensive modifications to the chassis or bodywork. The controls are also extremely simple. An on/off switch on the dashboard controls the petrol pump, and a hand-operated gas-valve regulates the flow of gas. The engine is started on petrol, with just enough suction to bring the gasifer into operation. Once the gas flow is adequate, the petrol pump is switched off. On a hill, the petrol pump is switched on and the gas valve switched off. The gasifier-powered vehicle is probably at too great a disadvantage to survive the snarling dogfights of traffic in metropolitan Manila, but it seems perfectly adequate for rural use.

Fuel is almost invariably charcoal made from wood or coconut shells. Wood chips and other materials may be used, but at this early stage GEMCOR advises users to build experience with charcoal only. This minimises the risk of tars clogging the filters or damaging the engine. People are not advised to aim for maximum fuel savings, but to concentrate on taking what is readily achievable without affecting efficiency.

The economics appear remarkably good. Gasifiers cost as little as $50/kW so the advantages of local manufacture and cheap labour rates are immediately apparent. Models imported from Europe tend to aim for higher levels of fuel saving, a wider range of feedstocks and a greater degree of automation in their operation. They are usually more complicated and expensive. Ex-factory prices for such gasifiers may be $400/kW.

The crucial issue seems to be technical abilities, and here the Filipinos score heavily. They are not just adept; they are irrepressible. They seem capable of making anything mechanical out of something else. Manila's public transport is based on a multitude of "jeepneys" the originals of which were surplus US army jeeps left after the war with Japan. Some are genuine 1940s vintage; others are locally-produced derivatives. Modern, unofficial vehicular hybrids such as the "Toyota-Benz" are not unknown. If it is mechanical, Filipinos seem to be able to manipulate it and adapt it. Gasifier technology is no problem.

The shadow over the future of the Philippines programme – and over any Third World gasifier programme – is the availability of fuel. Wood depletion and deforestation are already a problem in the country. Satellite pictures show that forests cover 30 per cent of the country; the government feels that 46 per cent coverage is a minimum, for economic and environmental reasons. Attempts are being made to link gasifier deployment with the planting of fast-growing species such as the ipil-ipil (*Leucaena leucocephala*), which may yield 25 tonnes of dry wood per hectare per year.

A key role in the tree planting is being played by the Farm Systems

Development Corporation. It encourages the formation of farmer cooperatives and advances them loans for irrigation pumps. Some 1800 of these have now been installed. Tree planting is also supported by a government campaign.

Some success is being reported, with a net gain in forest area of 34 000 hectares in 1980. But the basic hope is that market forces will drive wood and charcoal prices high enough to create an incentive for small-scale voluntary replanting and large-scale forestry projects.

Meanwhile in Brazil, a firm called Florestal Acesita is playing a key role in developing gasifier technology. This subsidiary of the steel manufacturer Acesita supplies its parent firm with charcoal for steel smelting. About 250 000 tonnes of charcoal per year are provided from Acesita's eucalyptus plantations. The average yield from these forests is about 20 m^3/ha/year, but yields of up to 50 per cent more are now being achieved with new plantations. Charcoal from planted wood costs around $60 per tonne to produce.

Florestal Acesita has maintained an interest in gasifier technology. The firm has acted as a clearing house for information on plantations, charcoal and gasifiers for years. It collaborates in prototype development and is testing a varity of gasifiers with its transport and planting fleets. Florestal intends to convert 176 large trucks to gasifier power.

Charcoal is used, almost exclusively, for vehicles and shaft-power applications in Brazil. This is the main reason for the extremely low cost of the country's gasifiers. The fuel is clean-burning which means the gas-cleaning system can be simple compared with systems burning a different biomass. No work on fuels other than charcoal appears to be in progress.

Available gasifiers are well-designed and economically competitive, but most have been on sale for 18 months or less. Private companies have carried out most of the development work, usually to meet specific markets. They are concerned with the problems of adapting designs to meet individual needs. To this end, considerable attention is being paid to establishing service and distribution networks: some are doing this in advance of their marketing efforts.

A couple of years observation will show what long-term impact gasifiers will have on Brazil. It is probable that some units on sale will find applications in other wood-rich developing countries.

This two-wheeled tractor in China runs on a mixture of biogas and diesel.

PART SIX

Power from muck

The biomass fuels we have concentrated on so far have been plants of one sort or another, either gathered from forests or cultivated as crops. But there is another important source of material from which we can extract energy: the millions of tonnes of organic waste that the world produces every day. And, unlike supplies of wood or land under cultivation, the amount of waste humans produce is increasing all the time. Britain alone throws away more than 8 million tonnes of organic waste a year, from domestic sewage, agriculture and industry. Modern agriculture, in particular, turns out vast amounts of slurry: a British dairy cow weighing 500 kg produces 41 litres of waste every day.

This biological material could be a vital source of energy. A report at the 1981 UN Conference on New and Renewable Sources of Energy (UNERG) estimated that cow dung and crop residues could supply 2 per cent of the world's energy. The amount of straw that Britain's farmers burn every year could make them self-sufficient in energy.

But as usual, the picture is not as simple as it looks. Unlike other forms of biomass fuel, you cannot simply put agricultural waste (except dried cow dung, an important fuel in the Indian sub-continent) into a cooker and set fire to it. Turning waste into energy usually means digesting it with bacteria to make biogas (a mixture of methane and carbon dioxide) and a harmless nutrient-rich sludge that can be used as fertiliser.

In theory, the process is straightforward. All you have to do is put the waste into a sealed container (a digester) and let anaerobic bacteria break it down. Gas can be tapped from the top, and sludge run off from the bottom as the pressure in the digester builds up. Digesters range in size from backyard units to vast "landfills" of buried urban sewage.

The idea of gas for free is an attractive one for both the developed and the developing world. In particular, these biogas plants could avoid the waste of burning dried dung for fuel, thereby depriving soil of fertiliser. As we have seen in the section on wood, biogas plants form an important part of developing countries' plans to become self-sufficient in energy. So far, the Chinese have taken the initiative, building some 7 million simple community-sized digesters providing gas for 35 million people.

But although biogas has enormous potential, it is not a panacea. Even the simplest plants, a hole in the ground lined with brick or cement, cost a lot of money by the standards of a Third World village. Nor are the technical problems solved. Fermentation has a habit of stopping in cold weather, and a certain skill is needed to supply the right balance of ingredients to keep the process going. Even when fermentation is taking place, a scum forms on the surface of the mixture, interfering with the production of gas. Lastly, there are problems with the gas itself. The output from small plants tends to be irregular, and biogas cookers have a habit of going out.

There are also sociological problems. In countries such as India it is the comparatively well off who are most likely to buy a biogas plant. As a result, cattle dung, formerly the fuel of the poor, becomes a market commodity out of reach of many people who used to depend on it.

These problems will be solved by a mixture of basic research and hard practical experience. On a basic level, as Eric Senior writes in this Part (p. 142) we need to know a lot more about the process of forming methane, or methanogenesis, and what factors help and hinder it. On a practical level, as the shorter articles culled from over the years show, there is still much to be learned about such factors as the optimum size of a digester and the practical problems of keeping it running.

First, however, we have to learn to look at waste as a resource, not as a problem. That is what the first article in this Part is about.

25

Waste not, want not

DAVID HUGHES and CLIVE JONES
20 March, 1975

Every year Britain throws away more than 8 million tonnes of organic waste. The authors argue that we would do better to recycle this refuse and effluent into useful products.

The main sources of organic waste in Britain, which we could exploit to save growing or importing food for ourselves or animals, are domestic garbage, sewage sludge, and farm wastes. By 1980 we will generate about 7 million tonnes of dry organic matter as domestic garbage, and 1–1.5 million tonnes as sewage sludge. These two are already collected by local authorities, so recycling them has a possible starting point; but it is difficult to say what might be done with farm wastes because they are dispersed, and some fulfil a useful role as fertiliser. As a crude indication of the scale of the resources lost by our present methods of waste disposal, the dry organic content in domestic garbage and sewage amounts to half the weight of cereal grown in this country.

Before we consider what might be done in detail, it will be useful to look at the principles underlying the use of wastes as raw materials for making new products. These are the same whatever the waste.

First, the products to be made from the waste, as well as the markets for it, must be identified. Secondly, especially where any new capital investment is concerned the future of any waste output must be established. It is no use establishing waste conversion on arisings that may disappear as industrial processes change.

Finally, the energy used in conversion must be minimised; this includes transport costs, and inputs of energy in the manufacture and maintenance of the product. Few such energy balances have yet been determined, but in general one would expect that microbiological conversion would need rather lower energy inputs than conventional methods of waste disposal.

In Cardiff, the biologists of the recycling group have been developing ideas and experimenting on the conversion of organic wastes to useful products using the above precepts. This has led us to suggest that organic wastes are best converted by the action of living organisms, from which, of course, they arose. At first our ideas evolved – perhaps somewhat naively – from ecological considerations as follows. Almost all organic substances arise by the reduction of carbon dioxide by energy from the sun: other elements, for instance nitrogen, phosphorus, and sulphur, are also incorporated.

Thus organic compounds represent a source of energy and, more important, an almost infinite range of chemical structures on which the use of specific natural products by man is based. In the complex chemical reactions from which all living things are produced, part of the organic material taken in by the creature is used as fuel for energy, while part is converted to other compounds using part of that energy. Dead creatures, and wastes that are produced from this process, are all re-used in this way: only gradually is the carbon re-oxidised to carbon dioxide. Rather than burning or burying organic wastes, as we do now, to let them decompose slowly, we propose using a selection of living creatures, including microbes, plants and animals, to convert wastes into useful products.

Technical and economic surveys of waste arisings and their potentiality for re-conversion have shown that in some cases, for instance domestic garbage and sewage, conversion on a sufficient scale into microbes to be used as a source of protein could make this country largely independent of imported fish meal and soya. These items cost £130 million a year in 1973, 8 per cent of our imports by value, or 4.3 per cent of our present balance of payments deficit. There would also be a promising export market in supplying other countries with the equipment and the essential know-how that is used in the conversion.

However, because wastes arise even from microbiological conversion, the linking of other biological engineering processes are proposed under the general title of bioplex, An Urban Farm System. These include the breeding of small animals, insects, and fish for the maximum product yield. These products include the living creatures themselves, which can be eaten, or alternatively processed by conventional abbatoirs into materials such as fat and fertiliser, for which there is already a market. This is not a new concept; many single systems have been used in the past. For instance, waste food and sewage went at one time, and still do in some countries, into moats or ponds where fish were cultivated. For decades wastes have been fermented to produce mushrooms and chemicals such as ethanol. However, modern biochemical engineering heralds a new dimension.

Domestic garbage is the principal source of organic wastes. The composition of garbage has been altering in most countries in the past decade so that there has been a substantial rise in the organic content (paper, food, cardboard, etc.). At the moment most domestic refuse is disposed of by filling existing holes such as quarries, but burial sites are in increasingly short supply. Baling reduces the space needed, but is expensive: the cost ranges between £1.50 and £2 a tonne. Composting before land fill or reclamation is not widely practised, although advocated by environmentalists. The costs of handling have risen steadily, to give a total of about £1.50–£1.75 a tonne. Incineration is more expensive even when part of the waste heat is recovered. Assuming a calorific value of 4500 Btu/lb, disposal by this means still costs £3 to £5 a tonne even without counting transport, because of the high capital cost of installations handling one or two thousand tonnes of rubbish a day.

In contrast, separating and recovering the useful parts of waste does not seem to be very much more expensive than throwing it away. Two main systems have been developed for separation. Dry separation by air elutriation recovers metals, glass, paper, food. Part of the organics may be composted, and part compacted and pelletised as animal feed, as in the Rome plant which can deal with 900 tonnes a day at a cost of £2.5/tonne. Wet processing by hydropulping and fibre chain plants such as the Black Clawson system give costs of £1.8 to £5.6 a tonne depending on the size and value of the re-covered pulp, metal and glass.

Another process, pyrolysis, cooks the wastes in an inert atmos-phere to yield gases and oils – all reduced carbon products. Except possibly in Belgium, no full-scale plant yet exists, nor, despite the oil crises have satisfactory outlets been developed for the products. One possible outlet is building board, but again, no full-scale plant yet exists.

The other waste source, sewage sludge, is normally settled to produce a primary sludge. It may then be treated by microbes to produce a secondary sludge which consists mainly of microbes and purified water. These sludges consist mainly of protein, about 35 per cent; fibre 15–30 per cent; fats 10–30 per cent; and

minerals, and thus can be considered equivalent to animal foodstuffs. At the moment a fifth of the national product is disposed at
sea, 40 per cent on agricultural land, and 40 per cent elsewhere. The
cost is about £11 a tonne of dry sludge.

Farm wastes consist largely of slurries and litter from stock and
poultry raising and residues from other farm processes, to which the
wastes from food industries must be added. The overall analysis is
much the same as sewage sludges: protein, fibre and fat, but the fibre
and moisture contents are more variable and depend on the source.
The growth of intensive farming is producing an increased quality
of concentrated wastes, whose costs of disposal are rising rapidly.
One might also mention forestry wastes here, especially because
only 18 to 20 per cent of a tree by weight is used for the final
product. The bark in particular is in bulk an embarrassment to
dispose of. So it appears that there is a large quantity of wastes
which could usefully be recovered and treated to produce a useful
variety of materials.

There are four factors to be considered in designing a bioplex
process using a particular animal to convert wastes to food:

(a) yield of body weight
(b) rate of conversion of waste
(c) marketability of animal product
(d) acceptability of the intensive farming process to the population
     close to the bioplex.

Whatever the animal chosen, whether it is a microbe or an ox, we
obviously want it to convert as large a proportion of the wastes to
food as possible. The rate of conversion is equally important
because the process will have to take in a large continuous stream of
waste: to keep the size of plant and amount of capital employed to
reasonable proportions, high conversion rates are necessary.
A plant to produce 1500 tonnes of protein a year would cost
between £2 and £5 million, and return between 25 and 30 per cent
on the capital employed at present prices. This return is for protein
alone, without counting other products such as starch or board
which might be derived.

Marketability of the final product is obviously very important.
This means an outlet for all the products at a fair price and is a major
restriction on the choice of animals. Most animals will be acceptable
if they are to be sold as feedstuffs for other animals, but if they are to
be sold directly for human consumption there will be obvious
problems of sales resistance. This could be a serious obstacle to the

concept. The acceptability will vary with the product. For example, pigs fed on fractions of domestic refuse and incorporated into pork products would meet less sales resistance than, say, ducks bred on sewage and sold for the table. Microbial protein produced from the cellulose fraction of refuse and used as protein enrichment in convenience foods would be more acceptable than rabbits sold as fresh meat. Market research will be necessary to aid the selection of animals – but if we are to feed the whole world, eating habits must change.

If we reflect on the changes that we have accepted in foods over the past decade we can see that our eating habits are not as conservative as some people would have us believe. However, if bioplex research does proceed, it must be supported at an early stage by a market research programme.

Another influence on the choice of animals will be the reaction of the people living close to the site. Microorganisms grown in large tanks or fermenters, or fish farms in deep siloes, will obviously be more acceptable to a large urban community than will intensive farms breeding goats or geese. But goats or geese will fit more readily into a rural bioplex than will the relatively complex machinery of fermentation. The different locations will therefore produce different products. This is a definite advantage in not saturating a market by over-production.

Therefore the schemes suggested for large urban complexes will mainly be fermentation processes. They will be like industrial processes already carried on in those areas, with the possible advantage that they produce no obnoxious fumes or dangerous effluents. The rural bioplex will be like an intensive farm.

Britain's organic wastes form a food resource equal in potential size to half our present production of cereals. The value of animal feed imports alone which might be saved by exploiting this resource amounts to some £130 million about half our balance of payments deficit in a pre-oil crisis year. A capital investment of £200 million or less would suffice for the necessary plant. Whether Britain is to become self-sufficient or not, it seems foolish to neglect this valuable source of food.

F

26

Will-o'-the-wisp goes to work

ERIC SENIOR

We have known for more than 100 years that waste substances digested in the absence of air generate useful methane gas. But we have been slow to exploit this resource. We need to know more about the basic microbiology to be able to do so.

Methane is one of the most promising alternative sources of energy. It is a good fuel and abundant: between 300 and 400 million tonnes of the gas goes into the atmosphere every year. Biogenic processes in animals – cows, termites, iguanas and humans – paddy fields, fresh and saltwater sediments, landfills and trees account for 60–70 per cent of the methane. The rest comes from emissions of natural gas, from burning fuel and biomass, and from coal mining. The atmospheric burden of methane is increasing at a rate of about 2 per cent per year, with only slight discrepancies between the northern and southern hemispheres.

For more than 100 years scientists have recognised the potential of anaerobic digestion (fermentation in the absence of air) to generate energy from waste such as manure and sludge. At the end of the 19th century the street lights of Exeter were occasionally fuelled with biogas from septic tanks, but since then most research on sewage has focused on treating effluents aerobically (by processes that depend on oxygen).

Although increases in the price of oil have rekindled interest in energy-conserving anaerobic treatments, it is still generally accepted that the greater benefit of the technology comes from treating effluent, and so protecting the environment, rather than from producing gas. For example, in Britain the reduced costs of discharging the end product account for 70 per cent of the savings made from digesting waste anaerobically.

But in other countries, fermenting domestic sludges and agricultural wastes to make methane is now quite widely acknowledged to be energy-efficient (Box 1). In China between 4 and 7 million family-sized anaerobic digesters turn out biogas for cooking and lighting.

However, the process has not yet received a similar endorsement as a way of treating industrial waste water. This is probably because of a legacy of previous problems. But more than 500 plants are now running at pilot or full scale in Europe, and are successfully treating, either alone or in combination with other processes, about 60 different feedstocks.

One case where the cost benefits of generating methane are anything but marginal is in the landfill bioreactor. This may be thought of as a multi-million cubic metre anaerobic digester. More than 25 million tonnes of waste in Britain goes into landfills every year. Disposing of it this way is up to 66 per cent cheaper than other methods such as incineration, or turning the waste into liquid fuel. Thus any gas that landfills generate must be a further bonus.

Box 1: The best of both worlds

There is little point in preaching to the Third World about the damage that burning cow dung does to the soil's fertility unless you can offer an alternative. The most appropriate is biogas, a mixture of methane and carbon dioxide, which is used extensively in China.

Biogas has often proved too complex and costly for small households but a relatively cheap and simple system has been developed in Thailand. It costs around £15. A polythene tube or bag is simply dug into a trench which can be made in different sizes for fermenting excreta from two to several cattle. Human excreta and vegetable waste can also be fermented. The gas produced from the excreta of two to three cows should provide enough fuel for a family to cook for 4 or 5 hours every day.

The slurry left after fermentation is effectively a new by-product, rich in microbial proteins (20–30 per cent in dry matter). The slurry has several uses: as a fertiliser, for feeding fish in fish ponds, or as a possible source of protein for calves or cows. The most important point is that the essential soil nutrients that the dung contains return to the land – either directly, or by enriching irrigation water – even after the carbon has been fermented to yield its heat energy.

More and more countries, including Britain, the US, Canada, Switzerland, Italy and West Germany, recover gas from landfills. Britain could raise 2 billion therms a year (60 000 megawatt-hours) from this source). But local authorities in Britain have been slow to exploit this potential, considering that the potential of landfill to generate methane was documented as long ago as 1934. The US has shown no such reluctance. Its Department of Energy estimates that the 20 billion cubic metres of methane that landfills generate every year could provide 1 per cent of the US's energy; the first site to recover gas, in the mid-1960s, was at Los Angeles, and the practice has now spread to the East Coast, to Fresh Kills, the world's largest landfill, on Staten Island, New York. Fresh Kills' full operating capacity will be 1.1 million cubic metres of raw gas per day, which will be upgraded to 500 000 cubic metres per day of 98 per cent pure methane.

Sites from which gas is commercially extracted vary both in area and in depth. But smaller landfills, with areas as little as 3 hectares, are now being considered. The amount of gas produced is difficult to predict and to monitor, but, conservative predictions suggest rates of between 3 and 37 litres of raw gas, containing about 55 per cent methane by volume, from each kilogram of refuse per year. Again, the concentration of methane in the raw gas will vary, reflecting both the characteristics of the landfill and the materials tipped.

Although we know that it is economically viable to abstract and purify gas from landfills of solid waste, scientists have made few definitive studies of the fundamental microbiology and biochemistry involved. It is therefore fortunate that comparable research in other areas of anaerobic biotechnology has been growing rapidly over the past 5 years. We now know much more about how methanogenic bacteria behave (Box 2). However, much work needs to be done, particularly on environmental variables and the kinetics of catabolic processes (the breakdown of complex molecules into simpler ones).

Techniques to extract gas from landfills usually involve sinking wells consisting of perforated pipes, and sucking out the gas with a compressor. As these techniques have developed, interest has focused on ways to optimise the generation of methane. We assume that all methanogens have similar environmental and physiological needs, regardless of the particular metabolic processes that predominate at particular sites. Although anoxic freshwater sediments and sludge digesters characteristically produce 60–70 per cent of methane from acetate and 30–40 per cent from carbon dioxide,

Box 2: A metabolic symphony

The conversion of complex organic polymers into simple molecules that include methane involves a series of stages as outlined in Box 3; hydrolysis to produce intermediates such as sugars; acidogenesis, producing complex acids from those intermediates; acetogenesis, converting the heavy organic acids into lighter, volatile fatty acids including acetic acid; and finally methanogenesis, the production of methane, primarily from acetic acid. These stages are effected by whole ecosystems of bacteria working in concert and in competition.

However, such a simple description does scant justice to the reality of methanogenesis within a complex substrate like a landfill. In general, any chemical reaction can be seen as the transfer of electrons from one molecule to another, and in a series of reactions an electron or group of electrons, is shuttled from along the sequence molecules. For reactions to take place at all the immediate surroundings must be able to donate or accept electrons. Indeed, the reality of methane production has been described as "a metabolic symphony of carbon and electron flow directed by the chemical composition of the initial organic electron donors and electron acceptors present in the environment".

Various common components of landfills enhance, inhibit, or divert the appropriate flow of electrons along the chain of reactions and thus enhance or inhibit the production of methane gas. For instance, if sulphate is present, bacteria that reduce sulphate to hydrogen sulphide will tend to divert both carbon and electrons away from the methanogenic bacteria, and so reduce methane production. The explanation for this can be clearly seen by considering the reduction of sulphate to hydrogen sulphide which is energetically more favourable than methane production for both hydrogen and acetate substrates. In addition, hydrogen sulphide at concentrations above 100–150 mg per litre is highly toxic to methanogens.

But sulphate is unlikely to have a major effect in landfill because the sulphur content is generally less than 0.2 per cent by weight. Indeed, anaerobic organisms such as methanogens need inorganic sulphur in order to synthesise protein (which contains small amounts of sulphur) and to make various other essential materials, including some that are involved in the transfer of electrons (such as ferrodoxins) and coenzyme M, an integral component of methyl group transfer reactions. Thus, for some microbial processes in landfills a shortage of sulphur could almost be limiting.

Taken all in all, landfill provides an extremely competitive and hostile environment for methanogens. They must compete for

BOX 2 *continued*

substrate with other bacteria that are intent on de-amination (removing the amine groups from amino acids) and reducing nitrates and sulphate. They may also be constrained by other environmental and nutritional factors. For instance, methanogens must have ammonia as a nitrogen source, and they perhaps form associations with bacteria that produce ammonia by degrading amino acids, as Japanese scientists have reported occurs in anaerobic digesters. However, methanogens do not need a great deal of nitrogen, and 0.5 per cent of nitrogen by weight, which is a typical concentration in refuse, should not be so low as to limit methane production.

However, methanogens also require iron, cobalt and nickel (as an essential component of the coenzyme F_{430}) as essential nutrients, and problems may arise when these are driven out of solution and thus made unavailable. Thus sulphide, even in non-toxic concentrations, may inhibit methanogenesis by precipitating essential trace metals. Similarly, heavy metals may or may not be toxic, depending on whether there are other molecules present that will bind (chelate) them, or cause them to precipitate.

Many other common components of landfill may also be toxic to methanogens. These include chlorinated hydrocarbons, detergents and fatty acids. Recent work in our laboratory has shown that phenol, catechol and quinol each significantly inhibit the formation of methane from veratric acid – one of the monomers (and a fairly minor one) from which lignin is compounded, the crucial ingredient of wood. However, phenol did not show this inhibitory effect when tested on a landfill model, with a large surface area of refuse; possibly because of physico-chemical interactions.

In general, the conditions that must be met if methane is to be generated by fermentation are a redox potential of less than −150 mV (and in some cases less than −330 mV)), warmth (that is, a mesophilic or thermophilic temperature regime) and, with only one exception, a near neutral pH. Neutral pH could be facilitated by ammonium generated from the breakdown of protein – which constitutes 2.5 per cent of the dry weight of domestic refuse – or by adding lime.

However, even when the methane is formed it is not sacrosanct. If it is not removed it inhibits further production; and it may then be re-oxidised anaerobically, by sulphate-reducing bacteria or methanogens themselves, or oxidised aerobically by methanotrophic bacteria.

Thus, efficient generation of methane depends in large part on removing it efficiently.

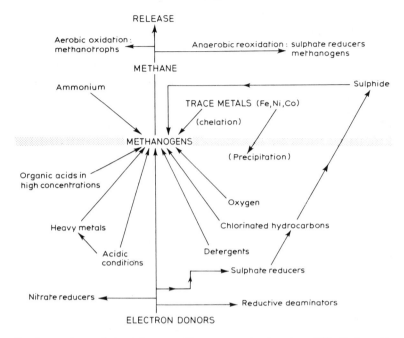

Factors at work inside a methane-producing landfill. Below the line are conditions that inhibit the production of methane; above the line what happens to the gas.

reduction work in our laboratory has shown that in an industrial and commercial refuse site another metabolic path is usually responsible. This involves methanol, which is formed when pectin, found in food wastes, breaks down.

One way to improve the generation of methane might be to saturate the refuse with water. Emplaced refuse often has a water content in excess of 30% by weight. Fermentation is best with a water content of 55–60 per cent. This approach has met with little enthusiasm, however, because it could pollute the groundwater. In any case, saturating the refuse might reduce methanogenesis, not least by introducing oxygen and so fostering aerobic catabolism (which produces carbon dioxide rather than methane). Perhaps a more environmentally acceptable method would be simply to recycle the liquid that leaches from the refuse back via a distributor, and thus to use the landfill as an anaerobic trickling filter.

Much work remains to be done before we can routinely establish high-yielding, stable, methane-producing fermentations. Similarly, we must pay close attention to developing efficient techniques of collecting and abstracting gas. These will need to overcome problems of gas migration, the problem of the safety of site workers, and that of replanting the site. Coal mining has provided much documentary evidence on the problems of uncontrolled gas production – in particular its movement, accumulation and accidental ignition. But until very recently we knew little about the effects of methane on vegetation, and assumed that plants on landfills died because bacteria around their roots used oxygen to metabolise methane, so depriving the plants. However, research at the University of Essex has shown that the picture is more complicated; species planted to reclaim sites often contain extremely high concentrations of iron. The explanation seems to be that as the gas migrates upward into the covering material it creates anaerobic conditions which reduce nitrate and sulphate, but more particularly increase concentrations of ferrous and manganous ions in the soil. This finding must cast some serious doubt on the common practice of grazing animals on landfill sites, especially where they are to be slaughtered for human consumption.

Efforts to optimise gas generation *in situ* could exacerbate these problems, and waste authorities may have to find alternatives. One possibility would be to generate methane by treating municipal refuse in an anaerobic "plug flow" system, although the quantities involved could cause problems. Alternatively, methane could be made by digesting leachate from landfills anaerobically. The amount and content of leachate is influenced by such variables as the geology, hydrology, hydrometeorology, topography, vegetation, the composition of refuse, and the physico-chemical factors linked to them.

Before any leachate becomes apparent the refuse field capacity must be exceeded as a result of liquid ingression and *in situ* generation by microbial activity. The chemical composition of the leachate usually reflects the age of the refuse. For example, leachate from material that has only recently been tipped contains high concentrations of fatty acids, particularly acetic, propionic and butyric acid. Conversely, once an active methane-producing fermentation has become established on the site, the organic content falls markedly.

Traditionally, waste authorities have regarded leachates as a problem, and have already developed efficient systems to collect leachates, and cheap processes, such as aerobic lagoons, to treat them. Although there are obvious advantages in digesting leachate anaerobically, the process has received little attention, partly because of the high capital costs. However, work in our laboratory has shown that leachates can generate high yields of methane.

An essential prerequisite for efficient anaerobic treatment of leachate is to inhibit methanogenesis in the site itself. This may be accomplished by rapidly eluting the refuse mass with water, producing high concentrations of fatty acids which lower the *p*H, and thus inhibit the production of methane.

Once we can identify the limiting factors in anaerobic digestion we can overcome them and begin to approach the theoretical maximum production of methane from domestic refuse, and even control the rate of output to match the energy produced to the demand. Because the quality of leachate varies with time, a series of digesters could be installed. Their number would at first reflect the

---

Box 3: *Search for a better fermenter*

Waste from the food processing industry poses just the kind of problems that technologies for producing methane gas can solve. The waste is plenteous, but is highly dilute: typically it contains about 5 grams of organic matter per litre. If the waste were more concentrated, it could be fed to yeasts or bacteria to make single-cell proteins, or upgraded to make industrial feedstock, or fermented to make ethyl alcohol for fuel. But when the material is dilute, the end product is liable to be too watered down.

What is needed is a useful end product, which methane is, that will come out of solution by itself. Methane does this most obligingly.

However, present-day methane digesters are not always as reliable as they should be. The basic problem, according to Dr Brian Kirsop of Britain's Food Research Institute, in Norwich, is that we do not understand well enough the microbiology of methane production. Already, however, scientists at the institute have devised a system that on a laboratory scale at least is more efficient than the best industrial fermenters, and plants are already afoot for a pilot scale industrial plant.

Methane can be produced from the whole gamut of organic material to be found in effluent from the food industry: carbohydrates, proteins and amino acids, and fatty acids. Whatever the starting point, however, the sequence of bacterial breakdown

BOX 3 *continued*

follows the same basic route. First, hydrolytic bacteria break down the polymers (proteins, peptides, oligosaccharides) to simpler molecules, such as sugars and amino acids. Then acidogenic bacteria break down these monomers into short-chain volatile fatty acids (VFAs) such as acetic, butyric, and proprionic; and along the way, hydrogen and carbon dioxide are generated.

Then comes the final transformation into methane, and this is the step that tends to create the bottleneck. The methane is produced in two ways. At least 20 species of bacteria are known to produce methane by combining carbon dioxide with hydrogen, and this kind of reaction accounts for about a third of the methane produced in a digester. But two-thirds of the methane is produced by reducing acetate (acetic acid) and this can be done by only three bacteria that belong to one of two genera: *Methanothrix* and *Methanosarcina*.

All the bacteria that produce methane, by whichever route, belong to an arcane group known as the archaebacteria. These organisms are similar to other bacteria in that they are prokaryotes (their DNA is not contained in a nucleus) but in other respects they are so different that they are given their own kingdom: that is, they are considered to be as different from other bacteria as plants are from animals.

Archaebacteria are strict anaerobes, highly averse to oxygen: they are thought to date from the time when the Earth was young, and its atmosphere contained no free oxygen. Nowadays they are to be found in swamps and in the bowels of animals, typically deriving energy either by methanogenesis by reducing the sulphur that wells from volcanic vents on the ocean bed. What counts for practical purposes, however, is that their biology is not well understood; and knowledge of their requirements must be improved if engineers are to improve significantly on the present generation of digesters.

Whatever it is being used for, whether generating methane from organic waste or turning out antibiotics, a digester must meet several objectives. Crucially, it must feed the microbes with a constant supply of substrate, provide ways of separating the desired end product, and maintain the population of microbes.

In practice, methane fermenters are of three basic kinds: those that keep the bacteria constantly in suspension, by continuous stirring; those in which the bacteria are allowed to form clumps and sink to the bottom of the tank to form a "sludge blanket"; and those in which the bacteria are held on the surface of some supporting medium, to form a "fixed film". All three kinds are in action in various places and on various scales around the world; but the kind under investigation at Norwich is the fixed film.

BOX 3 *continued*

Fixed film systems have many advantages. First, all the various kinds of bacteria involved in breaking down the organic waste are held closely together, so that the various intermediates they produce pass easily from one to the other. Then again, because they are stuck to a support, the bacteria are not washed away as the effluent passes over them – a problem with stirred tank digesters. Finally, if they are well designed, fixed film digesters are extremely efficient. In the system that Dr David Archer is working on at the Food Research Institute, the bacteria – hydrolytic, acidogenic, and methanogenic – are held on the surface of beads of pumice, held in a column. Effluent flows in at the bottom, and out again, devoid of its organic content, at the top; and methane bubbles off *en passant*, to be trapped in containers.

The polluting power of organic material is measured, for practical purposes, in terms of "chemical oxygen demand" (COD) per kilogram; the amount of oxygen required to render a kilogram's worth non-toxic. (The term may seem confusing in an anaerobic system, such as a methane digester, which does not involve oxygen, but it is a measure of potential; not necessarily of actuality). Engineers measure the efficiency of fermenters in terms of kg COD per cubic metre of fermenter per day. On these units (kg of $COD/m^3/day$) stirred tank reactors achieve a level of around 2, although contact digesters, which are an improved form of stirred tank digester, achieve levels of 4–10. Sludge blanket and fixed film systems in present use achieve around 10–20 with easily degradable waters. The laboratory-scale model at Norwich achieves 26 kg of $COD/m^3/day$. This is using effluent that is relatively easily broken down; but even with extremely difficult material (with a high content of sulphate) the model achieved a figure in excess of 10.

But what counts at this stage is that the system demonstrably works, and that the basic microbiology that underlies its performance is being improved. As the basic science improves so the engineering can follow. To the food industry, methane production is primarily a way of disposing of what would otherwise be a pollutant; a very desirable way, as all the carbon is converted to innocuous gases, either methane or carbon dioxide.

In other circumstances, as on isolated farms and villages, where there is organic waste to spare and a shortage of fuel, methane is desirable as fuel. Thus, although different forms of waste present different engineering problems, everyone should theoretically benefit from advancements in understanding underlying principles of methane production.

size of a site, but would be reduced as time went on and the concentration of leachate dropped. This would minimise the need to store leachate and gas in tanks, and would allow digesters to be moved from site to site as needed. As an alternative, centralised digesters could serve a number of sites, although it could prove too expensive to transport leachate.

Up to now, most gas taken out of landfills has either been up-graded to the standards of natural gas and injected into pipelines, or has been sold as low-grade energy to consumers such as brickworks or commercial greenhouses.

A big problem in Britain is that industries are often not close enough to a landfill; a pipeline longer than about 700 metres is too costly. Perhaps new companies should be given financial incentives to use methane-producing sites.

Alternatively, authorities could consider generating electricity from methane, or liquefying the gas for fuel. Liquefaction seems the more attractive, because most sites need a lot of vehicle fuel, and it is comparatively cheap to transport. Besides, if electricity is generated from landfill methane, a significant proportion of the potential energy is lost, and there is no point in generating electricity from methane if it is cheaper to buy it from the grid.

Despite its obvious attractions, we must regard generating methane from landfill leachate as a short-term objective, because the project has a relatively low market value compared to other products of digesters, such as reduced organic chemicals. However, although landfills represent a reservoir of untapped potential for making more profitable chemicals and chemical feedstocks, there will be no motive to realise this potential until the rising costs of petrochemicals make it attractive.

27

Manure into gas

"SCIENCE IN INDUSTRY"
26 January, 1958

At first, reports on biogas plants were tinged with optimism. But as experience built up, problems began to appear, as items 27–29 show.

Research into the microbial formation of methane, which has been in progress at the National Chemical Laboratory for some time, is now reported from the Haifa Technion, Israel. A pilot plant for the production of methane from cow manure, constructed from locally made concrete and metal parts, has been demonstrated to a gathering of farmers, scientists and engineers. It comprises two fermentation tanks and a gas holder, based on a patented design by two Frenchmen.

Four months of experiments suggest that the plant can yield 3400 m³ a year. The gas contains 55 per cent methane. The gas from the demonstration unit will fuel the steam boiler of the dairy at the Agricultural Research Station, Bet Dagon. Alternatively, it could be piped into the mains supply, and thus reduce fuel imports. Development of larger plants is being actively pursued. After the fermentation process, the manure can still be used as fertiliser.

28

Porker power

"ENERGY FILE"
28 November, 1974

"Farming uses less than 3 per cent of Britain's fuel supplies to produce over 50 per cent of our food", says the Ministry of Agriculture, Fisheries and Food in a not very useful pamphlet on fuel on the farm (*Focus on fuel*). While the ministry tells farmers to block up the draughts in their animal sheds, L. John Fry offers a more positive contribution to the farmer's energy balance. He overcame the problem of what to do with the two tonnes of manure produced daily by a thousand pigs on his South African farm: he turned it into methane.

Twenty years ago Fry started developing methane digesters. His book on the way in which he overcame the problems thrown up during his development programme has just been published in Britain.

Methane production sounds nice in theory, but like all such ideas when you try to do it countless problems crop up. With no background of recognised expertise, and no "literature" of straightforward "how to do it" details, the would-be methane generator could soon find himself knee deep in manure, wondering why there is no sign of gas.

Fry's book should reduce the hazards of digester construction. He describes digesters as small as an oil drum, and big enough to process 350 tonnes of pig manure a year. He also tells you how to store and use the methane; and he suggests some safety measures you can take to reduce the risk of blowing yourself up by igniting a dangerous mixture of methane and air.

Fry's book goes through all the difficulties he experienced when building his series of methane makers. The "scum problem" was probably the hardest to overcome. After a digester has been operating for a while, a thick layer of solid scum forms on the surface. This gradually reduces the efficiency of methane production until eventually it stops altogether.

Few of the alternative technology articles on methane production devote much space to the difficulties as well as the benefits of methane production. Fry's book shows that methane production is not something to embark upon lightly, but it shows the would-be gas maker that he is not alone when he comes up against those difficulties that inevitably occur with such systems.

Gobar gas plants: beyond farmers' means?

29

Small biogas plants are not so beautiful

"TECHNOLOGY"

6 January, 1977

For those parts of the world most in need of alternative energy sources such as methane from animal and vegetable wastes, the household-scale fermenter is not practical because it costs too much – whoever is paying. This was one conclusion from an international conference on biogas, organised by the Canadian government's development and research group.

A fermenter to provide a family's methane for heat, cooking and perhaps lighting, at a rate of 6–10 m³ of gas per day, fermented from four cows worth of dung plus agricultural wastes, costs around £250. That is far beyond the means of an Indian or Philippine farmer who really needs the gas.

A community fermenter, providing about 100 m³ of gas at a cost of around £1000 would be within the means of most of the people, although not all, in need. Producing gas on a much larger scale begins to lose the advantages of village-scale technology.

The conference, in Colombo, identified clear needs for research. Too many fermenters have been introduced in development projects without enough attention either to local needs or to efficient engineering. Research projects should look at the design of burners to cope with damp gas, the design of pipes for distributing gas locally, cheap water traps to replace expensive steel gasometers; and the extent to which waste should be chopped up before putting in the fermenter.

30

Cooking up biogas for tea in Sri Lanka

"TECHNOLOGY"
11 October, 1979

Sri Lanka's experience with biogas as part of an alternative energy strategy was typical of many developing countries.

The development of new energy sources in Sri Lanka has become a matter of extreme urgency. The government's policies for boosting rapid industrialisation and the expanding rural population are creating an energy shortage, despite President J. R. Jayawardene's decision to accelerate the Mahaveli hydroelectric scheme. The future holds only increased dependence on imported coal from India and on expensive nuclear power, unless Sri Lanka can develop alternative sources, such as solar and wind but most importantly biogas, to supply the rural areas, including the vast tea estates.

Christian Aid, with the promise of future help from Britain's Overseas Development Administration, has just agreed to finance five pilot biogas plants which, if successful, could be copied across the country to ease the energy shortage in rural areas. Although the sum of money involved is a small £10 000 the project's success could seed similar projects in Thailand and other parts of the Third World.

Unfortunately, the government is ignoring plans for these trial plants, designed by Sri Lankan scientists and suited to the country's rural areas, in favour of a much larger, Western-designed biogas plant, which some believe is inappropriate and costly. The United Nations Environment Programme financed the development of this much more ambitious design.

Most villagers and plantation workers derive their energy for cooking and lighting from firewood and kerosene. Firewood has become so scarce in some areas that villagers are reported to be burning branches from avocado and mango trees during the off-season. And kerosene is subject to an enormous government subsidy variously estimated before the most recent oil price rise at between two and three times the cost to the consumer. Sooner or later the government will have to pass this subsidy on, and events in 1973/74 have already amply demonstrated how vulnerable rural populations are to even the smallest rise in fuel prices.

Because supplies of hydroelectric power were once considered to be virtually inexhaustible, governments have paid very little attention to the development of the renewable alternatives. Dutch experts have developed an inexpensive multiblade windmill which can be used to pump water in rural areas and the Appropriate Technology Group of Sri Lanka is working with the Intermediate Technology Development Group in London on a cyclone-proof design. Solar devices are still much too expensive to be of any significance except as a means of producing hot water for hospitals and hotels. So biogas is the only alternative energy source capable of reducing the effects of imminent fuel price increases on the rural population.

Biogas – a highly combustible mixture of methane and carbon-dioxide – can be used for cooking, lighting, refrigeration, and as a substitute for diesel in a small engine. It is not always recognised that from an overall economic point of view the most valuable feature of biogas generation is the fertiliser sludge which remains after the gas has gone. It is rich in phosphorus, nitrogen and potassium, and no longer contains undigested weed seeds present in animal dung prior to anaerobic decomposition. The fertiliser slurry may be applied directly to crops, and can also be turned into algae to feed fish by exposing it to sunlight.

A survey carried out earlier this year discovered that, despite a widespread feeling that biogas should be developed extensively, only two units are operating with any degree of success. The first of these, at Gannoruwa near Kandy, has been designed and built by members of the government's Department of Agriculture, and Peradeniya University. The second is a much larger unit constructed by the United Nations at Pattiyapola in the extreme south.

The Gannoruwa biogas plant is a scaled-up version of what is basically an Indian design suitable for between 6 and 10 people and costing about £140. Cow dung from three or four cows is sufficient to provide gas for cooking and lighting, and the unit can easily be expanded to incorporate algae and fish ponds. As a fuel, its 90 per cent conversion rate makes it highly efficient. At Gannorwa the gas also fuels a water pump to irrigate nearby crops and

Cooking with biogas in Sri Lanka.

vegetables. The criticism that biogas plants deprive the small Third World farmer of a source of income from selling cow dung does not apply in Sri Lanka. There it is spread directly onto the soil as fertiliser and not sold.

The Pattiyapola biogas unit, built under the auspices of the United Nations Environment Programme, is part of a larger scheme which converts biogas, wind and solar energy into electricity. The combined electricity from these three sources is stored in a battery bank and transformed into alternating current by an alternator. The resulting total of about 60 000 kWh, is more than enough to light and pump drinking water into 200 homes at Pattiyapola.

Although the Pattiyapola scheme has undergone several modifications since its inception, it none the less incorporates many inappropriate features, and it is unfortunate that the government is committed to imitating it. Even allowing for overheads which are unavoidable in any experimental project, it is grossly extravagant to spend $233 500 on the energy needs of a mere 200 homes. The solar generators are particularly costly and the wind turbines are of a novel and untried design.

Another problem is that the originators of the scheme were led to believe that sunshine and wind would complement each other throughout the year in the dry southern zones. In fact, this is not the case for two months each year. Cattle in the area are free-grazing, making the collection of cow dung difficult. It would appear that the scheme's authors did not set foot on Sri Lankan soil until after the project had been sanctioned.

31

Hills are alive with vine energy

"TECHNOLOGY"
10 July, 1980

Cow dung is not the only waste material that can produce heat. Research in Austria suggests that waste grapes could play their part in saving fossil fuels.

Austrian researchers think that grape marc – the seeds, skins and stems of the fruit left over from wine making – could be an important source of energy and fertiliser.

Eighty-five per cent of the world's grape production is turned into wine and juices. But 10 per cent of the grapes – 40 million tonnes – goes to waste after it has been pressed. Each tonne of dry marc can provide about 10 000 MJ of energy and so has a fuel value between that of brown and black coal, and above that of wood. But unlike wood, grape marc cannot be burnt directly because it has a high ash content.

In a research programme funded jointly by the Federal Ministry of Science and Research and the Austrian firm of Vogel and Noot, scientists have identified between 20 and 30 yeasts, fungi and other microorganisms which digest waste from wine pressing. In the early stages of digestion, when there is plenty of air available in the mounds of grape marc, sugar in the skins and stems left over from grape pressing, is converted into energy.

While the stems and skins are rotting, the grape seeds remain unaffected. But when the seeds are ground down, microorganisms that can tolerate temperatures above 45°C attack the seeds which release carbon dioxide and water vapour. Placed in buckets in greenhouses, the seed mash helps the growth of horticultural crops by releasing more carbon dioxide during the day, and extra heat at night.

Grape marc contains a high ratio of carbon to nitrogen and so can promote the growth of microbes. It also contains a lot of phosphorus, another nutrient important to microbial growth.

One difficulty was how to use effectively the large amounts of heat produced as the grape marc rots – temperatures within the mounds of waste can reach 80°C. Eventually, Graefe installed small, thin radiators in the middle of the mounds of grape marc. He says that it takes microorganisms 5 months to consume a 5 m^3 block of grape marc: this should provide enough energy for a small room or stable to be warmed throughout a long winter.

PART SEVEN

The gasohol gamble

The most ambitious, sophisticated – and controversial – biomass energy projects to date involve making alcohol from crops to eke out supplies of petrol for cars. The fuel is ethanol (ethyl alcohol), better known as the vital ingredient in alcoholic drinks. Its history as a fuel is a long one: Henry Ford built cars that ran on mixtures of alcohol and petrol, and during the depression of the 1930s farmers in the American Midwest promoted alcohol distilled from corn as a fuel to keep farm prices up.

However, cheap oil eclipsed the fuel alcohol industry, as it did many other alternative energies, and it was not until the oil shock that planners started looking seriously at alcohol again.

As every home brewer knows, a vast range of plants can be fermented to produce alcohol. The first stage for most crops is to break down the starch with enzymes, into sugar. Of course this is not necessary if the raw material is already in this form – sugar cane, for example. Yeast then converts the glucose into ethanol plus carbon dioxide ($C_6H_{12}O_6 \rightarrow 2C_2H_5OH + 2CO_2$). The resulting solution should be as strong as beer, containing about 9 per cent alcohol.

To make pure alcohol, it is necessary to distil this product, and this is one of the most controversial stages of the operation. There is not much point in burning oil to distil alcohol that contains less energy than you have put in. As we shall see, at least one plant, in Kenya, has failed because of its "negative energy balance".

However the most controversial aspect of "gasohol", as the mixture is known in the US, is the amount of land needed to produce enough alcohol to make an impact on a country's oil bill. Brazil, the country with the most ambitious plans (p. 167) will have to turn 2 per cent of its land over to sugar cane to become self-sufficient in fuel. That is half the area of all cultivated land now in the country. Apart from the fact that food could be grown on that land, there are also doubts about the ecological effects of turning such vast tracts over to single crops.

Despite the doubts, the US has pressed ahead with its gasohol plans, although the recent oil glut has removed much of the incentive. Here, the feedstock is grain, of which US farmers produce a huge surplus. The target is to replace 10 per cent of petrol with alcohol by 1990, but this seems unlikely to be met unless there is another sharp rise in oil prices. Even if this happens, many people will doubt the wisdom of devoting a hungry world's grain to keeping America's gas-guzzlers on the roads.

32

Brazil avoids hiccups with alcoholic car fuels

TREVOR LONES

18 September, 1980

Brazil leads the world in the technology of making alcohol from sugar and cassava and using it in road vehicles. It even exports know-how to the US. Its ambitious plans to reduce the country's dependence on petroleum spirit by making even more alcohol should keep it ahead.

To the casual observer, the petrol gushing out of the pumps at Brazil's petrol stations may look just like that served in any other part of the globe. But it is in fact very different, containing 20 per cent by volume of dehydrated ethyl alcohol, and is the result of the world's first large-scale coordinated programme to substitute petroleum with fuel alcohol. The Brazilian government's Proalcool project began in 1975, and by 1978 had saved more than 1.6 million litres of petrol by substituting alcohol, and had held petrol consumed by road transport, previously increasing by around 10 per cent per year, virtually static. The government's target is to reach a production figure of 10.7 billion litres of alcohol a year by 1985 which, according to best estimates, at that point could be substituting some 40 per cent of the country's petrol consumption.

Through the distribution efforts of Brazil's national oil company Petrobras, virtually all of the country's 27 states and territories now sell the alcohol/petrol blend. This has been made possible by the fact that neither cars nor pumps need to be modified and for this reason the blend has been maintained at a maximum of 20 per cent alcohol. The government originally aimed for a 25 per cent mixture, but severe corrosion of petrol pumps ruled this out.

Petrol/alcohol blends, however, are merely a step along the road toward the use of pure alcohol by all Brazilian motorists. Some 205 pure alcohol pumps in the main industrial and commercial centres serve the 250 000 pure alcohol cars already using the highways.

Most of these vehicles are specially built, but about 50 000 are standard models converted at a cost of around $400 each. By the end of 1980, there should be no fewer than 350 000 factory-built models on the road. If the National Association of Automobile Manufacturers gets its way, the production of petrol-powered vehicles could end by 1981–82 and the factories would start to make millions of new all-alcohol cars.

The incentives to the motorists are real enough. Pump prices for blended petrol were hiked for the second time in a month in June and now stand at cruzeiros (Cr$)34.50 a litre – £1.08 per British gallon. The pump price for alcohol stands at Cr$18 per litre and even with a 25 per cent loss of efficiency due to the lower calorific value of ethyl alcohol, the economics speak for themselves, especially as fuel prices will continue to rise.

The reasons why Brazil should be pushing flat out to find petroleum substitutes are not hard to find. Despite massive and continuing efforts to find oil onshore and offshore (the whole of Brazil's sedimentary basin is being opened up to exploration by foreign oil firms) the country supplies itself with only around 18 per cent of its oil needs. The rest has to be imported at an annual cost of around US$10 billion, eating up half of the country's total foreign

Cars fueled on alcohol in Brazil.

earnings. Brazil also has an overseas debt that will reach $50 billion – the largest in the developing world – so interest payments are about $9 billion a year. When combined with the oil bill, this threatens to outstrip foreign earnings altogether. Economists estimate that, taking current oil consumption into account and based on the assumption that prices will rise by 20 per cent each year, Brazil's oil bill in 1985 will be running at $25 billion. If this happens it is doubtful whether there will be enough national income to pay for the oil it needs to keep its industries going.

The choice of alcohol as a fuel substitute was an easy one to make. Ethanol/petroleum mixtures in various proportions have been used from time to time in many countries and Brazil was somewhat of a pioneer in the concept. The technology was developed back in 1931 and stimulated further during the Second World War when oil supplies to Brazil dropped almost to nothing. Alcohol has the advantage over petrol that it is non-polluting when burnt in cars and can also be produced from regenerable feedstocks such as sugar cane. To cultivate sugar cane successfully, three things are necessary – sunshine, water and large tracts of agricultural land, and Brazil has plenty of all three.

But it was not until 1975 that a Proalcool-type project became economically feasible, compared with importing oil. In the early stages of the programme, the spare capacity of existing sugar mills allowed Brazil to operate on the marginal economies of adjacent distilleries – making alcohol as a by-product of sugar. Oil prices have now grown to levels where it is economically worthwhile to process sugar cane in special agro-industrial plants designed to produce only alcohol – autonomous distilleries.

The total amount of alcohol that Brazil's existing sugar mills can produce as a by-product is a maximum of 5–6 billion litres a year. Increasing it would reduce the 7.5 million tonnes of domestic sugar produced – of which 5.5 million tonnes are consumed by Brazilians. The whole question of how much sugar is exported from existing plantations is subject to the wildly fluctuating world markets. Ten years ago for instance, the price of sugar was $1500 per tonne, last year it was $150 a tonne while earlier this year the price was $650 per tonne and rising. How much sugar is available to convert to alcohol therefore fluctuates with market prices and cannot be guaranteed.

As an interesting aside, processing costs are obviously much lower if liquid sugar materials such as molasses are used as the basic feedstock. At one stage, Brazil diverted molasses into alcohol production, an action which pushed the world price of molasses to an all-time high. At that stage it was obviously more attractive to sell the molasses directly, buy in oil and still have some cash left over.

The petrochemical industries use similar arguments over the use of anhydrous (free of water) alcohol from sugar cane as a source of fuel. Anhydrous alcohol, the industry says, is a pure chemical and as such should not be mixed with relatively impure petroleum but rather used more economically as a feedstock to replace naphtha in the manufacture of petroleum derivatives. Ethanol can be readily converted to ethylene which is the primary building block for a wide range of common products such as polyethylene and polyvinyl-chloride. It is paradoxical that the import of naphtha petroleum feedstock forms a significant fraction of Brazil's total oil imports.

Nevertheless, Brazil is determined to push on with the Proalcool programme. Industry has been fairly successful in building distilleries next to existing sugar cane plants and most of this year's 4 billion litres (60 000 barrels of oil per day equivalent) will come from this type of unit. But to achieve the 1985 target of 10.7 billion litres per year its farmers will need to plant 1.5 million hectares of cane – an area almost the size of Brazil's existing plantations which account for 15 per cent of the world's crop – and the construction of some 300 autonomous agro-distilleries. A typical plant that can produce 120 000 litres per day costs in the region of Cr$5–15 million (US$100 000–300 000) for the land and planting, plus some Cr$10–12 million each for the distillery and establishing the infrastructure.

The government has already approved more than 281 projects, with the potential to produce 8 billion litres a year. But so far more than half of these projects have not got off the ground, despite the Cr$54.9 billion (US$1.094 billion) allocated in the 1980 budget to alcohol production. The 1980 Proalcool budget provides: Cr$22.34 billion for industrial distilleries; Cr$10 billion for agricultural distilleries; Cr$18 billion for agricultural and plant-ing; Cr$2.68 billion for grants; and Cr$1.89 billion for miscellaneous demands. The total is Cr$54.91 billion.

Many observers doubt if the government is capable of reaching its target of 10.7 billion litres by 1985 as this would require the distil-leries to produce an additional 120 cubic metres of alcohol every week between now and then. By the turn of the century, however, few doubt that given the steady application of cash, production could be hitting the equivalent of 1 million barrels of oil per day without adversely affecting the food crop.

Any home wine or beer maker is familiar with the treatment accorded to any carbohydrate (sugar or starch) feedstock during fermentation. In the case of sugar cane, the cane is crushed in huge mangles, the resulting sugar solution is diluted, yeast culture is added and the liquor kept in huge vats at ambient temperatures until fermentation is complete – usually when the alcohol level has reached approximately 9 per cent by volume, when the yeast is killed. The yeast culture currently used – *Sacchraromyces cerevisiae* – acts on the glucose molecules producing two molecules of ethyl alcohol and two molecules of carbon dioxide for each molecule of glucose. Formaldehydes, glycerol and some superior alcohols are also produced but these comprise no more than 0.2 per cent by volume and are left behind in the final residue.

The fermented mass is then mechanically filtered to remove most of the solids and the resultant liquid submitted to ordinary fractional distillation where hydrated ethyl alcohol is taken off at 78°C. Until recently the fractionating columns were made of expensive imported stainless steel, but under the Proalcool programme, new plants are constructed from locally-produced, ordinary carbon steel, the pH factor of the fermented solution first being neutralised to prevent corrosion.

The hydrated alcohol so produced contains 96 per cent of ethanol at maximum and can be used directly as a fuel substitute in all-alcohol vehicles where a water content of up to 10 per cent is perfectly acceptable. For blending with petrol, however, the alcohol has to be completely anhydrous and this is achieved by azeotropic distillation with benzene or hexane. Benzene is more readily available and thus most commonly used. The benzene associates with the water and is left behind at 69°C, when it can be recycled for further use.

One of the great benefits of using sugar cane as a basic feedstock is that there is a net energy gain. The cane not only provides the basic fermentable sugars but also the bagasse fuel for running the distillation process. In fact more bagasse fuel is produced than is necessary and about a third of it can be accumulated during the cropping period to power the distilleries when the crop is out of season to process other feedstocks such as stored syrup, molasses, cassava extracts, sweet potato, sugar beet and so on, which may have a negative overall energy gain.

Sugar cane takes 1½ years to grow to maturity and can be cut and cropped only twice in its lifetime – a process which begins in the north of the country in April and lasts around 7 months, and in May in the south where cropping takes 5–6 months. A second and final cropping takes place 6 months later.

Brazil has been criticised for the crude and inefficient way it handles the fermentation and distillation of its carbohydrate feedstocks. Certainly alcohol is being produced now in a hundred or more small distilleries, some of which are extremely antiquated. During a recent visit of the US secretary of industry, a group of Brazilian businessmen suggested that the two countries participate in a joint venture, with the US supplying capital investment and technical know-how for the fermentation process which apparently could be made more efficient with a chemical not available in Brazil. There is also a West German process which deals more efficiently with the sugar cane by dissolving out the sugar in water with suitable additives. Brazil has one such plant working.

Brazil's best plant geneticists are making an all-out effort to improve the amount of sugar produced per tonne of sugar cane. But even now a mere 3 per cent of available agricultural land in Brazil would, if given over exclusively to sugar cane to produce alcohol, make the country self-sufficient in energy.

Sugar cane is just one of the available feedstocks, and in the future planners are likely to give more emphasis to a root crop which has provided the staple diet for the Brazilian poor from time immemorial – manioc or cassava. There are 200 different varieties of cassava in Brazil and the government is now selecting five of these – which are highly resistant to disease and have a high carbohydrate content in the form of starch – for alcohol feedstock. On average, the cassava root contains 30 per cent starch, but using variants with a 35 per cent content, it has been possible to produce 80 tonnes of root per hectare – the break-even is 15 tonnes per hectare.

Unlike sugar cane the cassava will grow almost anywhere, even in the poorest soil, provided the pH is to its liking. Under the Proalcool plan the poor soil of the Cerrado region of the country, which occupies 25 per cent of Brazil's total 8.5 million km^2 of agricultural land, should be planted with cassava in rotation with soya bean. Cassava prepares the ground for soya beans and by rotation, the latter crop has been increased by 50 per cent.

At the moment very little alcohol is produced from cassava. There is one pilot plant currently being operated by Petrobras at Curvelo, producing 60 000 litres per day, and a second will come on stream this December.

Every part of the cassava root finds a use. The peel and residue

The cassava plant.

from fermentation are used as high protein cattle feed, the carbon dioxide gas from the fermenters make dry ice and carbonated drinks. The residue which is too aqueous to burn direct, and comprises 12 : 1 by volume of resultant alcohol, is also concentrated into a paste which can be used to make lubricating oils or as a feedstock for producing methane gas. The foliage of the plant can be dried and burned, or fed to cattle. There is in fact a net overall energy gain although there is not enough bagasse to drive distilleries. The root can, however, be dried, stored and treated in sugar cane refineries during the out of season period using excess sugar cane bagasse, resulting in continuous alcohol production.

Brazil is the largest producer of cassava, or manioc, in the world. Easy harvest and culture on all types of soil and reasonable productivity has encouraged small farmers right across Brazil to grow it. Now, the government has started to encourage these small farmers to use part of their cassava crop to make alcohol to power tractors and farm machinery. Under the guidance of the National Institute of Technology, some of these small farmers are now producing up to 1000 litres of ethyl alcohol a day – although how much is consumed by farm equipment and how much by farmers is hard to say.

The disadvantage of cassava compared with sugar cane, molasses or syrup is that the starch content has to be converted to sugar before it can be fermented – all starch feedstocks suffer from this problem. In the case of cassava, the root is peeled, crushed and mixed with water to form a crude mash with a fixed starch content. The mash is steamed to break down the vegetable cell walls and enzymes added in the form of alpha-amylase and glucoamylase. These enzymes convert the starch into glucose which can then be fermented and distilled in a conventional fashion.

It is theoretically possible to obtain 185 litres of ethyl alcohol from one tonne of "saccharified" cassava. Pilot plants have already reached 178 litres to the tonne.

Liquid fuel has become firmly established as the most acceptable and, in many cases, the most economical source of energy for road vehicles. With its convenient properties and almost universal acceptance, alcohol has been considered as an alternative fuel as long as there have been internal combustion engines. Having taken the lead in setting up the first large-scale alcohol fuel programme, Brazil is now poised to start exporting its technology. Saudi Arabia and Iraq have expressed interest in investing in the Proalcool scheme primarily with the intention of forming an alco-chemical complex, or for exporting alcohol to similar industries. Sudan, with its fast growing sugar industry and a transport system incapable of handling its molasses by-product, is considering a power alcohol programme rather than dumping the molasses in the desert. Thailand, with a wide variety of fermentable crops – tapioca, maize, rice and sugar cane – is considering channelling its surplus into alcohol production instead of selling on a weak market. Columbia, Cuba and the Philippines are looking at cane juice, Papua New Guinea is considering cassava and the Ivory Coast is pondering on using molasses. The US is likely to go ahead with a corn-based alcohol programme, supplemented perhaps with sugar cane from Louisiana and Puerto Rico.

Brazil is also selling technology in the one area in which it has undisputed experience and success – blending alcohol with petrol for use in the motor car. The US, for all of its advanced technology, has turned to Brazil for help in blending the fuels and solving carburation problems. Blending technology has also been sold to Costa Rica. Zannini SA Equipamentos Pesados, one of Brazil's biggest sugar and alcohol distillery equipment manufacturers, has won export orders of US$8.5 million to Mexico, Panama, Costa Rica, Argentina, the Philippines and Kenya. Another $29 million of foreign orders are in the pipeline, Zanini says, and feelers are out for contracts valued at a further $42 million.

As an amusing footnote, the Brazilian government has decided to blend the pure alcohol sold at the pumps with 3 per cent petrol. This has nothing at all to do with improving the blend or the efficiency of the cars. It is simply that the national drink of the Brazilian is composed of virtually pure ethyl alcohol mixed with ice and fruit juices. The alcohol pumps were apparently fuelling not only the cars, but the drivers as well.

33

How Brazil's gasohol scheme backfired

"THIS WEEK"
16 July, 1981

But there was another side to the Proalcool scheme. In the run-up to the United Nations' Conference on New and Renewable Energies, *New Scientist* reported:

One of Brazil's leading environmentalists has made a blistering attack on his country's policy of fuelling cars with alcohol to save oil. Jose Lutzenberger, an agronomist, says that the programme, called Proalcool, is destroying unique ecosystems, exploiting working people, misusing land and encouraging massive spraying with potentially dangerous pesticides.

The Proalcool programme began in 1975. The government said that it had saved the country 1.3 million litres of petrol by 1978. Its target is to produce 10.7 million litres of alcohol a year by 1985. Official estimates say that this will reduce the nation's oil consumption by 40 per cent, but Lutzenberger says the real figure is nearer to 20 per cent. In the first stage of the programme, sugar is the main raw material, but in the longer term the plan is to develop technology to rely on cassava, a root crop that grows profusely in Brazil.

Lutzenberger says that the real story of Proalcool is one of multinational companies manipulating Brazil's government to make inappropriate use of arable land. For a start, he says, Proalcool was originally thought up by the sugar barons in the north-east region of the country. The federal government became enthusiastic once the motor industry had been persuaded to drop its opposition. It has been offering cheap loans for setting up alcohol distilleries – but only to large plants.

Secondly, although Brazil depends on trucks for commercial transport, alcohol is no substitute for diesel fuel. Lutzenberger said: "We could give up our cars . . . we cannot give up our trucks. We would starve." And many lorry drivers would be ruined if the government insisted they use gasohol. He says that this "small calamity" pales into insignificance compared with the environmental, social and economic problems that Proalcool poses for Brazilians. For example, one alcohol business, in southern Mato Grosso do Sul state, is growing 1700 square kilometres of sugar cane to produce 1–1.5 million litres of alcohol per day – less than 1 per cent of national fuel needs.

There would be no objection to huge plantations in Brazil's 8.5 million km^2 if they were on the several million hectares of land abandoned by smallholders who have gone to the cities. But instead, the Fezenda Bodoquena enterprise of bankers and sugar barons from Sao Paulo has levelled 1700 hectares of a "unique ecosystem of 'cerradao' forest", with a technique the US army developed during the Vietnam war – pulling the trees down with a huge chain dragged by two tractors. The timber is burnt. "In the Amazon, similar alcohol farms are now razing enormous tracts of forest for cassava," Lutzenberger said. He also alleges that the banned defoliant Agent Orange is being used in Brazil.

Lutzenberger claims that "Proalcool will be a boon to the pesticide industry . . . hypermonoculture makes the use of soluble chemical fertilisers and agrichemical poisons inevitable." Brazil is already the third largest consumer of agricultural chemicals in the world.

The main by-product of these vast alcohol plants is "a highly concentrated organic soup", which could be raw material for fertiliser, biogas or even cattle feed. Instead it is dumped into rivers. Lutzenberger singled out one offender – a plant owned by Brazil's minister of the environment.

But perhaps the worst consequence of Proalcool will be its effect on working people, by displacing small farmers to the cities. In the North East, where big landowners have always controlled the land, preventing "a healthy, locally adapted peasant culture from developing", working people have, in desperation, moved to the big cities such as Rio de Janeiro, Sao Paulo and Belo Horizonte, where, Lutzenberger says, "the filthy slums . . . overflow with Nordestinos". He predicts that Brazil's alcohol programme will spread this problem of migration and its attendant social ills to the rest of the country.

Lutzenberger says he is not against alcohol fuel *per se* – only the way that Brazil is developing it: "There is no energy crisis for a country the size of Brazil, but we do have a crisis in our technological, social and political models."

34

Carter slips gasohol under Reagan's driving seat

"TECHNOLOGY"
1 January, 1981

Meanwhile the US was following Brazil's example.

Planning to leave his mark on the US's new administration, departing President Jimmy Carter has announced plans to ensure that all federal-owned vehicles will run on gasohol fuel this year – or at least the transport already converted to run on gasohol will.

The scheme is another try by the government to boost the flagging fortunes of the industry that manufactures alcohol for fuel. The fledgling industry needs an increase in demand for its product. And it is not particular about who buys it.

A big section of the federal fleet of cars, lorries, vans and buses already runs on a mixture of petrol and ethanol – it consumes well over one million litres of alcohol every month. For over a year, Carter has been urging federal agencies, which run a total of 450 000 vehicles, not counting military equipment, to convert to gasohol. Last June, Congress rewarded his efforts by passing the Energy Security Act, which requires federal authorities to use ethanol whenever it is feasible and the fuel is available.

A recent analysis of ethanol production from biomass by the Department of Energy (DOE) has thrown a more favourable light on the problem of "net energy balance", the cost accounting procedure that has consistently shown that producing alcohol for fuel consumes more energy than it saves. One answer to saving oil and natural gas, the country's scarcest fuels, is to use coal as a process fuel in a more efficient plant: one that, for instance, conserves energy by recycling waste heat. Dr Edward Blum, formerly director of the DOE's Office of Advanced Technology, comments that distilleries offer much room for improvement in fuel efficiency. They have always depended on a big mark-up on alcoholic spirits to cover their high running costs.

The DOE also observes that more fuel is consumed than produced when ethanol is made from biomass: "The purpose of the biomass-alcohol process, however, is to take a form of energy with low utility and convert it into a premium fuel", says the DOE's analysis. With this approach, a nation can conserve its scarcer fuels and achieve a "net fuel gain". This arises when drivers substitute ethanol for petrol in their cars so that overall savings in a premium fuel in the tank exceed the spending on a premium fuel to run the process that makes ethanol for formulations of gasohol.

Using coal as a process fuel and corn as a feedstock, and operating in the most efficient available factories, alcohol manufacturers can save 0.83 litres of crude oil every time their customers replace a litre of petrol with a litre of ethanol in their vehicles. Should gasohol – about 10 per cent ethanol, the rest petrol – offer more kilometres to the litre than straight petrol, even more crude can be kept in the ground.

In January 1980, Carter set a target for ethanol production in the US of 500 million gallons by 1982. But while several hundred thousand more alcohol-guzzling cars may provide some cheer for the distillers, their numbers are unlikely to sweep the alcohol-for-gasohol industry into the big time.

35

American automobiles turn to alcohol made from corn

"THIS WEEK"

15 January, 1981

After 18 months of studying the prospects of an alcohol-based fuel industry in the US, a special presidential commission is waxing bullish over the possibility of running a big fleet of cars on pure alcohol, without dislocating food prices. Should the US's distilling capacity reach several billion litres within the next decade, however, the price of corn will rise in the US, along with other food prices. That would be an occasion for joy among American farmers, but would be less than desirable for consumers around the world.

A session at the American Association for the Advancement of Science on the "fuel–food conflict" previewed results of the forth-coming report by the National Alcohol Fuels Commission. The commission's analysis will greet a new president already committed to the American farmer. Ronald Reagan has repeatedly criticised the outgoing Carter administration's embargo on grain sales to the Soviet Union as unfair to American agriculture.

While the commission played down the effect on prices of the massive alcohol fuels campaign that is now in the works, economists at the meeting agreed that even modest production goals by the end of the decade could change worldwide habits of livestock feeding and human eating.

Congress has set a goal of displacing 10 per cent of all petrol consumption with alcohol based on biomass by 1990. That would mean building an ethanol capacity of more than 40 billion litres – about 100 times as much as the US makes today.

Most of the feedstock for current production is corn. But most of the corn the US exports goes to developed countries, not to the poorer nations, which receive mainly wheat and rice. Furthermore, the US stores more than 1 billion bushels of corn as surplus to maintain prices. Therefore, argues the commission's representative

Marilyn Herman, the poor need not starve to fuel America's weekend vacations, nor will induced shortages send the price of corn soaring.

Nor does making ethanol from corn destroy its value as feed. While the starch content is used up in producing alcohol, distilling 10 litres of alcohol from a bushel (35 litres) creates more than 30 kg of high-protein by-product and corn germ, and 34 kg of carbon dioxide.

Tests on livestock at several universities in the US show that the by-products can be added to low-value products, such as corn stover, to produce feed that is as bulky as the original corn feed, and produces leaner, better-quality beef.

Should the US attempt to meet its 1990 alcohol-fuel goals using only corn, however, the price of corn would probably double, according to Milton David, an agricultural analyst from Kansas. David observes that for each percentage point that alcohol replaces petrol at the pump, another 5 per cent of the nation's corn production would have to be diverted from food to fuel. That means Congress's 1990 goal would swing about half of America's corn supply into the petrol tank.

At some point below that level of production, however, price fluctuations can be kept to an "acceptable" range while still reducing the country's dependence on foreign oil. To defuse the controversy about sacrifices made at the dinner table for four-wheeled mobility, attention must be turned to available biomass that does not feed humans – wood, forage crops and crop residues that are now wasted.

Depending solely on corn, argues Wallace Tyner of Indiana's Purdue University, would shift the worldwide production of beef to chicken, a more efficient converter of corn to protein. Such a shift could result even at a biomass production level of 16 billion litres per year in the US, because the US provides almost half the corn traded internationally. To avoid havoc on the international grain market, the Alcohol Fuel Commission reports, technology for converting cellulose plant fibre to alcohol must be pursued. Cellulose feedstock can be obtained from such diverse sources as corn cobs, tree farms, waste from sawmills, crop residues and even garbage.

The commission will recommend that the US government cajole or subsidise the automobile industry into making cars that run on pure alcohol, which would then provide the impetus for increased ethanol production. Although the likelihood is low that the hard-pressed American auto industry will take such an experimental leap, reports from investment analysis in New York and Washington DC point to a rising interest in alcohol fuels.

In fact, large brokerage houses are now taking on more analysts skilled in distilling and related technologies to direct the flow of new private money toward the best ventures. And the US Department of Energy, in processing requests for federal subsidies to alternative fuel technologies, reports a massive influx of alcohol fuel projects in search of funding.

36

Fuel crops threaten the Third World's food

"THIS WEEK"

20 March, 1980

But some reports were more sceptical.

The developed world's hunt for new sources of fuel, which now includes alcohol derived from food crops, is so serious a threat to food production that it threatens to drive apart the rich and poor as perhaps nothing else has ever done before. "The price of oil may soon set the price of food", says Lester Brown, president of Washington's Worldwatch Institute in Worldwatch Paper 35, *Food or Fuel: New Competition for the World's Cropland.*

"As countries turn to alcohol distilled from agricultural commodities as a source of fuel for automobiles, more and more farmers will have a choice of producing food for people or fuel for automobiles. They are likely to produce whichever is more profitable. The stage is set for direct competition between the affluent minority, who own the world's automobiles and the poorest segments of humanity, for whom getting enough food to stay alive is already a struggle."

Brazil and the United States have both announced programmes to convert crops into alcohol, and other major food-exporting countries such as Australia, New Zealand and South Africa are also considering the possibilities of "fuel crops". A car engine can easily run on a gasoline/alcohol mixture containing as much as 10 per cent alcohol, and the technology for converting plant materials to ethanol is already widely available.

Brown's report points out that "the cost of producing ethanol is determined by such factors as which commodity is used as the feedstock, the effect of weather and location on the crop-yield, the value of any by-products, the size of the distillery and the type of fuel it uses and the subsidies available for alcohol production and use".

However, feedstock alone accounts for half the price of ethanol. But Brown says that as gasoline prices rise, gasohol (a mixture of ethanol and gasoline) will become more expensive and gasohol manufacturers will be able to afford more expensive feedstocks. In this way "the price of oil may soon set the price of food".

Brown paints a gloomy picture of the outcome of this competition for the world's agricultural land. Brazil's alcohol fuel programme, launched in 1975, is based on sugar cane, the most efficient of all energy crops, yielding 65 per cent more alcohol than maize, the feedstock the United States favours. Brown says "government plans to produce 10.7 million litres of alcohol by 1985 will require nearly three million hectares of sugar cane, the equivalent of 10 per cent of Brazil's cropland". Yet a study in 1975 showed that only one-third of all Brazilians were eating a sufficiently nourishing diet. Brown suggests that instead of sugar cane Brazil should turn to cassava for its gasohol, which can be planted by small landowners on marginal land in the least developed regions of the country.

The impact of the American push toward gasohol on the world's food supplies will be more far-reaching. In January 1980, President Carter announced that the US would aim to produce 2275 million litres of fuel ethanol by 1981 and 910 thousand million litres by the mid-1980s. Brown points out that the goal of 2275 million litres will require the output of almost a million hectares of farmland or five million tonnes of maize, which represents approximately 5 per cent of the US's projected maize exports in 1980.

The world's hungry millions are directly competing for this land. Brown says that since 1950 the amount of land devoted to growing food crops has increased only one-fifth as fast as the world's population.

But Brown's message is not entirely pessimistic. He says: "a carefully designed alcohol fuel programme that gave farmers first priority in the use of ethanol for tractors, farm trucks and irrigation pumps would help ensure future food supplies when oil begins to dwindle. Such an emphasis would be a major step toward the creation of a sustainable food production system and of a sustainable society. "

37

Alcohol in cars gives a bitter energy balance

"TECHNOLOGY"
20 March, 1980

The fact that modern farming consumes a lot of energy makes it difficult to see how agriculture can contribute much to the US's energy needs by producing alcohol. Scientists at the Coastal Ecology Laboratory at Louisiana State University have completed a "net energy analysis" of alcohol production from sugar cane. They calculated how much energy is consumed in the fuel and chemicals needed to grow sugar cane and added this to estimates of how much energy would be needed to turn the crop into alcohol.

Alcohol production is seen as an important energy option for the US because the oil companies can add alcohol made from plant products to petrol as an "extender". Gasohol, as this fuel is known in the US, can contain as much as 20 per cent alcohol before any changes have to be made to a car's engine.

The scientists found that the net energy balance – the ratio of the energy output to the energy input – depends upon what fuel is used in the plant that turns the sugar cane into alcohol. If fossil fuels power the production plant then the net energy balance is 0.9; that is, less energy comes out of the plant than goes in as fuel. The picture looks brighter if the waste from the sugar cane plantations, known as bagasse, fuels steam-raising boilers. Under these conditions the system can produce 1.8 times as much energy as it consumes, but some of it is in the form of steam.

If bagasse provides only as much energy as is required to turn the cane into alcohol, the energy output/input ratio is 1.5. But that is not the way in which the US expects to fuel its alcohol plants. According to Hopkins and Day, plants under design in Louisiana will use an approximately 50:50 mixture of bagasse and fossil fuel to fire the industrial apparatus. The resultant net energy balance would then be 1.2.

"Alcohol from crops produced in excess of food requirements will not make a significant contribution to national energy needs," the researchers conclude.

38

Third World energy plant bugged by hiccups

"TECHNOLOGY"

27 August, 1981

Meanwhile, Africa was also having problems with the new technology.

On the shores of Lake Victoria, engineers are building an alcohol plant that should by rights be a showcase of how the Third World can exploit new and renewable energy. Sadly, the reverse is the case, but for different reasons than one would expect from a troubled project in the industrialised world.

The plant, based on well-estabished Swiss technology, is designed to turn molasses from sugar factories within the Nyanza region of Kenya into alcohol and several other useful products. Current plans are for the plant, at Kisumu, to produce 20 million litres of fuel alcohol (to be blended with petrol in the ratio of 15 per cent to 85 per cent); 3000 tonnes of citric acid for the soft drink industry, 1800 tonnes of baker's yeast, and 2160 tonnes of vinegar. It will also produce sulphuric acid, fertiliser, gypsum and carbon dioxide. Sophisticated waste water treatment – accounting for 10 per cent of the capital cost of the plant – will also produce enough methane gas to supply 50 per cent of the plant's requirement for fuel.

On paper it sounds wonderful, but in practice it exposes all the weaknesses of developing industrialised technologies in a country whose people and markets have different priorities. First, the plant has been delayed by 16 months in the past two years, at a cost of about £325 000 per month. One delay was forced on the Kenya Chemical and Food Corporation when the main contractor went broke. The second was due to the government delaying cash grants because it was uncertain that the plant would ever be profitable. The plant will now cost £52.5 million (at present estimates), which is nearly double the original figure.

When it comes on stream, perhaps in the early months of next year, the plant will need feedstocks of molasses from several sugar factories within the radius of about 100 km. In theory, the plant will give a net energy gain of 5.2 billion kJ per year. However, Kenya's electricity authorities have not yet agreed to increase the electrical capacity of the local grid, so the plant will have to start up with diesel generators. Methane from the waste water treatment will not be available either in the initial stages, so diesel will have to provide power for process heat as well.

In the meantime, Kenya has two other alcohol plants under construction. One will produce another 20 million litres of alcohol per year by mid-1982, while the other, due to be completed in 1985, is three times bigger. Even 40 million litres of alcohol, blended in the ratio 15:85 with petrol, will exceed Kenya's current consumption of fuel. Although demand is rising at 7 per cent each year, the 60 million litres of alcohol that the Riana plant will produce in 1985 will swamp the market. So far, the Kenyans have not worked out to whom they will sell the excess alcohol and at what price. And the multinational oil companies which supply Kenya's petroleum do not yet support the blend of alcohol and petrol.

Despite these problems the plant is generally accepted, even by its critics, as being necessary. Matthew Midika, a local MP, says that the additional employment – 850 people directly and another 3500 indirectly – is essential in Kisumu, Kenya's equivalent of a grey area. Also, only 50 per cent of the present volume of molasses is exported, a figure that is likely to drop as the world price for sugar falls. The remainder was being dumped in the rivers, "killing my people", Midika says, so he welcomes any increase in its use.

Perhaps the main reason for continuing with the Kisumu plant, Midika says, is that it will show the world that the Kenyans are prepared to finish a project despite the difficulties. He says that a number of other high technology projects have been abandoned, with consequent damage to Kenya's credibility. Gilmore Bell, the managing director of Eximcorp, a Panamanian company that took over management of the plant in 1980, agrees up to a point. Ideally, he believes, the plant should be scrapped. The original feasibility study was poor (it only covered production of alcohol), the increases in cost have been devastating, the raw material will be imported from uneconomically large distances and the payback time will be at best nine years. Nonetheless, he concedes that in the environment of a developing country such as Kenya, circumstances are very different. He is open about the plant's weaknesses and the causes of them, but believes that the present targets will be met.

In the end, if plans do not go further awry, the losers will be those Nyanza locals who now use the waste molasses illegally to produce a local brand of gin known as Changaa.

39

Sticky end for molasses

"THIS WEEK"
10 June, 1982

The crunch for the Kisumu plant came in 1982.

The Kenyan government has abandoned a scheme for making fuel alcohol from molasses because it has realised that the system could never be economic. It has already spent about £40 million on the project – and the unfinished plant stands as a white elephant on the shore of Lake Victoria. To finish it would cost just £350 000.

The government is now busy disclaiming any responsibility for the 51 per cent stake in the project, which has been at the centre of a political row since its costs rose threefold. The plant, based on proven Swiss technology, was designed to process molasses from sugar factories in the Nyanza region of Kenya. It was to produce 20 million litres of fuel alcohol a year, to be blended with petrol in the ratio of 15 per cent alcohol to 85 per cent petrol.

But government sources say there is no chance of completing the project – and that the scheme would never have been given the green light if Kenya's Ministry of Energy had been in existence when the proposals were made.

The reason for the project's poor economics is the need for expensive fuel oil for heating and distillation. Although the government has made no formal statement about the future of the plant, it is considering salvaging the distillation equipment and installing it at one of Kenya's main sugar factories. This would theoretically allow the plant to be fuelled with baggasse (the residue after sugar cane is crushed).

According to government sources, the Israeli contractor, Solel Boneh, has walked off the site in protest about not being paid, and the Swiss plant designer, Process Engineering, has gone into liquidation.

40

New engine revives plans for alcohol fuel

"TECHNOLOGY"
14 October, 1982

Could a specially designed engine make gasohol more popular? The ever optimistic Brazilians decided to give it a try.

A car company in Brazil claims to have solved the big problem with running cars on alcohol: corrosion in the carburettor. The innovation could pull the country's fuel-alcohol programme out of the slump of the past two years. Ford Brazil was more cautious than its competitors in embracing the country's move towards alcohol fuel, but now believes its development of a carburettor bowl made of zinc coated with nickel has made up for the delay.

The corrosion problem forced Ford to design an alcohol engine from scratch, rather than adapting an existing petrol engine. This involves adjusting or altering at least 200 different parts.

The man responsible for designing Ford's alcohol engines is Fernando Barata de Paula Pinto. "We did have a lot of trouble with the carburettor," he admits, "nickel coating on the zinc has proved to be the answer. It's expensive, but it's paying off."

The new engine, which took de Paulo Pinto and his team at Sao Bernardo, the industrial suburb of Sao Paulo, just eight months to design, will run on either petrol or alcohol. The company says the decision to go for a dual-fuel engine was more to do with the Latin American export market rather than any doubts about alcohol's future as a fuel.

"The efficiency of our new engine will show a sizeable improvement on current alcohol engines," de Paulo Pinto said. He was not prepared to say what sizeable means, but the smile on the face of this earnest and rather serious engineer suggested that it would be significant.

Ethanol has about two thirds of the energy content of petrol. But its higher octane rating, which allows a higher compression ratio, and the lower fuel-to-air ratio combine to make an ethanol engine only 20 per cent less efficient than its petrol counterpart.

But it may take more than improvements in engine performance to end the reservations Brazilians have about running their cars on alcohol.

Brazil produces only about one fifth of the 1.8 million barrels of oil that it consumes every day. The 1973 oil crisis encouraged the government, like most others, to look at ways of reducing imports. Alcohol was one way, and the ambitious "Proalcool" programme was launched. The government set a target of making 10.7 billion litres of alcohol every year by the end of 1985, and the motor industry was to have put 1.2 million alcohol-powered cars on the roads by the end of this year.

The car companies responded well – if anything too well. At one time more than half the new cars sold had alcohol engines. Monthly sales peaked at 53 500 in November 1980. The government encouraged the move by subsidising prices, and allowing sales of alcohol at weekends, when petrol stations were closed.

But the rush to meet demand left the manufacturers too little time to refine the necessary engine modifications. People found that their new cars would not start, corroded quickly, and used too much fuel. The gap between the prices of petrol and alcohol began to narrow, and government officials were giving conflicting information about alcohol's price and availability.

On top of that, the recession began to bite, and the motor industry became badly depressed. Sales of all cars plummeted. In 1981 the industry sold just 2700 alcohol vehicles each month.

Meanwhile, the National Alcohol Commission trimmed its production targets for fuel: it will not reach its 1985 target until 1987. And controversy continues to surround environmental concerns such as the uprooting of food crops to make way for sugar cane.

On the credit side, there is no doubt that the Proalcool programme has reduced Brazil's oil consumption considerably. Dr Omer Mont'Alegro of the Sugar and Alcohol Institute, says oil consumption has dropped by 20 per cent over the past 3 years.

So although the alcohol programme is behind schedule, there is little doubt about its future. The World Bank has loaned the project $250 million, and Brazil is already selling its knowhow to the US. Optimistic assessments say that Brazil will save $600 million a year by the end of 1985.

41

Three years on: gasohol is respectable

T. P. LYONS

12 April, 1984

While gasohol or fuel alcohol is not receiving as much press as it did three years ago and may thus appear to the outsider to be a dead issue, it is, in fact, very much alive and, considering the age of the industry, very healthy. In 1981, the US Department of Energy indicated that by the mid 1980s the country would need as many as 3000 distilleries to meet its target. This figure was at that time considered to be on the conservative side if the small farmer operating his own unit for his immediate fuel needs was to have any impact.

The situation today is far from that anticipated by the Department of the Environment. Operating plants number between 50 and 100, compared to the forecast of 370. Many plants have either failed to start up or have started up and closed down. Small-scale plants on farms which flourished in the early stages have now very much disappeared. The reasons for the failure of so many have been threefold:

• *Emphasis on capital costs.* In the early stages of the alcohol industry, emphasis was placed on the capital cost of building a plant as opposed to the operating costs of running a plant. Many would-be producers based their final decision on which plant to purchase on the capital cost per output. Quotations ranged from 60 cents/gallon (US), to $3.50/gallon (US). Failure to do feasibility studies and to check the reputation and track record of the equipment makes led to poorly designed plants and operating costs in excess of the potential sale price of the alcohol. Economies in the capital cost were made by substituting inferior equipment made from mild steel, with the consequence of massive corrosion problems, in some cases within months.

• *Lack of operating capital*, coupled with the emphasis on total capital costs, poor financial planning led to inadequate operating capital. The plants that were capable of operating were unable to weather the first six months until money started coming in. Development of local markets for alcohol took more than was anticipated, leading to even greater requirements for operating capital. Many companies were forced into bankruptcy, making the industry a risky one for suppliers.

• *Lack of technical know-how*. In almost every case, plants were started up by people who had no experience in operating plants or whose experience was in a non-related industry. This lack of experience existed on the part of both the owners of the plants and the manufacturers of the plants. High temperatures during fermentation led to inactivation of yeast. Insufficient education on cleaning procedures led to infection and losses of yield. In many cases, on the advice of suppliers, details of the process were too complicated to run from a practical stand point requiring, in some cases, at least two unnecessary pH adjustments from operators who did not understand what pH meant in the first place.

Unlike in South America, where the alcohol programmes were controlled rigidly by the government, in the US free market forces were allowed to operate and in the burst of enthusiasm, many gross errors were made.

Farm production, which, if handled properly, could have contributed significant levels of alcohol, failed because of its inability to produce a saleable alcohol (anhydrous), in addition to fuel that could be used on the farm ($160-190°$ proof). The small capacity of the plants and poor design made the plants not energy independent as forecasted but energy dependent because of the inordinate amount of time needed to produce small quantities of alcohol.

Despite these difficulties, demand for gasohol has been increasing steadily. In 1981, total sales were close to 4 billion litres. This number translates into 400 million litres of fermentation ethanol for fuel. Sales figures for 1983 improved very much on this, and nearly 20 billion litres of gasohol were sold in 1983. This represents about $3-4$ per cent of total gasoline sold. Higher gasohol sales in 1983 have come partly from improved supplies, with more ethanol plants on stream, but also by a change in marketing strategy that was led by Archer Daniels Midland and Texaco. In many cases, the product is no longer sold as gasohol but rather as "ethanol-enhanced" motor fuel or "super unleaded with ethanol". The main thrust is octane enhancement. All Texaco stations, for example in Kentucky, sell "super unleaded" or octane boosted and a closer scrutiny of the pump shows it to contain "ethyl acohol". The subtlety of the change can often be seen in the attitude of the local petrol station managers who carry gasohol and promote the use of ethyl alcohol. The typical blend is 10 per cent, although in certain areas up to 20 per cent has been used without any modifications to cars.

PART EIGHT

The future: beyond biomass?

The ultimate goal of scientists working on biomass energy is one day to free the mechanism of photosynthesis from the burden of producing plant tissue and harness it purely to producing energy. One possibility would be to split water cheaply into hydrogen and oxygen, a goal that has for practical purposes eluded scientists working with conventional chemistry. Hydrogen could be the basis of the so-called "hydrogen economy", an all-purpose fuel as efficient and versatile as hydrocarbons, but renewable and non-polluting. In the first article of this Part, David Hall and colleagues describe some pioneer work on photobiological processes. Lionel Milgrom brings the fast-moving story up to date on p. 213.

Another possibility, though even more far-fetched, is the development of a "biological battery". This idea arises from the discovery that certain purple bacteria which live in the Dead Sea develop a minute potential difference when the Sun shines on them.

The goals are exciting but as the three articles in this section show, progress is uncertain and fraught with difficulties. The only thing we can be certain of is that nature has many secrets yet.

42

Plant power fuels hydrogen production

DAVID HALL, MICHAEL ADAMS,
PAUL GISBY and KRISHNA RAO
10 April, 1980

Solar energy keeps plants growing. By mimicking natural processes we may be able to turn water into hydrogen and oxygen using sunlight as the energy source. This would provide a clean, renewable fuel that would last virtually indefinitely.

An exciting development in renewable-energy research was the discovery, about 6 years ago, that light could split water into hydrogen and oxygen using membranes containing chlorophyll with enzymes added to the water to act as catalysts. This was the first step toward the development of a "biological" system that can turn solar energy into a useful fuel, hydrogen. In those early days the process continued for only minutes before the system "died" – a reflection of the instability of biological components when removed from their natural environment. Over the past few years laboratories in the US and Europe, including ours at King's College, London, have prolonged the life of the systems more than 20-fold so that they carry on producing hydrogen for 10 hours or more. At the same time the rate of hydrogen production has gone up by an order of magnitude.

It is not difficult to turn water into hydrogen and oxygen. An electric current applied to water through electrodes also splits water. This is the well-known process of electrolysis which is employed in industry and in chemical laboratories. However, electrolysis is not a way to make a storable energy source because valuable electricity – itself usually derived from stored energy, coal, oil, and gas – is used, and there are heavy energy losses in converting fossil fuels into electricity and then hydrogen.

Clearly solar energy can be turned into electricity and that can, in turn, split water by electrolysis, but this process is inefficient; it is probably better to try to get sunlight to do the job directly without having to go through an intermediate stage of electricity generation. Photochemists and photobiologists in industry and universities are interested in this problem, not just because of the energy opportunities that solar hydrogen production from water could open up but also because it involves interesting basic research. Photolysis has unique attributes that are unmatched by any other known energy system: the substrate (water) is abundant; the energy source (sunlight) is effectively unlimited; the product (hydrogen) can be stored and is non-polluting; and the process is completely renewable because when the hydrogen is "consumed" the substrate, water, is regenerated. Another attraction of this system is that it operates at normal ambient temperatures and does not involve toxic intermediates.

There are two approaches to splitting water with solar energy. Biological systems are very different from the photochemical hydrogen production processes known so far, which have their advocates in the research community. Photobiological systems use water as the ultimate source of the electrons that combine with the water's hydrogen ions to produce, and give off, hydrogen atoms. All the photochemical processes reported so far are only partial systems – visible light has not yet been shown to split water catalytically to provide electrons for hydrogen production. Unlike photobiological systems, whose reactions take place in membranes, photochemical processes tried so far have employed mixtures of chemical compounds. As yet researchers in this area have had only limited success, with one unconfirmed report of hydrogen and oxygen production in water containing four different chemicals.

The problems in constructing a stable biochemical system that will function for years rather than hours are enormous and may never be solved. However, if we could understand how the biological processes work and then possibly mimic them by constructing a completely synthetic system we might eventually be able to harness the solar energy that is available in temperate and hot climates. Such systems would use light at all intensities and temperatures; an attribute they would share with electricity-generating photovoltaic solar cells. Only the intensity of the light would determine the rate of hydrogen production. However, photobiological hydrogen production is still at the experimental stage and there has to be sustained and diverse research before it is ready for further development. Having said that, any "breakthrough", especially in the water-splitting reaction by photobiology or photochemistry, could quickly change our idea of the future practicability of such a system.

If we want to set up a photobiological process that produces hydrogen we do away with all the enzymes involved in the fixation of carbon dioxide and use only that part of photosynthesis which generates an energy potential close to that characteristic of the hydrogen electrode. Hydrogen production with the aid of chloroplast membranes was first demonstrated in the early 1960s, when hydrogen was produced in a system containing chloroplasts isolated from spinach, artificial electron donors instead of water, and bacterial extracts containing hydrogenase, the hydrogen-activating enzyme. The reaction involved was:

$$\begin{matrix} \text{Electron}^{e^-} \\ \text{donor} \end{matrix} \rightarrow \text{Photosystem I}^{e^-} \rightarrow \begin{matrix} \text{Electron}^{e^-} \\ \text{carrier} \end{matrix} \xrightarrow[\text{H}^+]{} \text{Hydrogenase} \rightarrow \text{H}_2$$

The hydrogenase enzyme accepts electrons from ferredoxin, the electron carrier. Ferredoxin and hydrogenase are both proteins which contain clusters of iron-sulphur ions as their "active centres". However, in these experiments the capacity of the chloroplasts to split water was deliberately suppressed and organic compounds were the electron donors for hydrogen production, that is only Photosystem I was used.

It was not until the early 1970s that researchers started to explore the possibility of producing hydrogen by splitting water, that is Photosystem I plus Photosystem II. Researchers in the US showed that on irradiation with visible light, in the absence of added electron donors, spinach chloroplasts and bacterial extracts containing hydrogenase would evolve hydrogen if either ferredoxin or viologen dyes were added as electron carriers.

The hydrogenase usually used, that from the bacterium *Clostridium*, is very sensitive to oxygen, which causes it to deteriorate rapidly. Hence the reaction is usually carried out in an atmosphere of nitrogen, with scavengers (glucose oxidase plus glucose) added to remove the oxygen which is evolved from water. With such a mixture under optimal conditions of pH (about 7) and temperature (about 25°C) we have produced hydrogen at a rate of about 50 micromoles per hour per milligram of chlorophyll. If scaled up this would correspond to a production rate of about 1 litre of hydrogen per hour per gram of chlorophyll. Our system produces hydrogen for about six hours.

Although researchers have made significant progress in improving this process, the biggest problem we must overcome is the inherent instability of the biological components involved. The instability of the isolated chloroplasts, particularly under continuous illumination in the presence of oxygen, is the main limiting factor. However, more stable chloroplasts have been isolated – for example, from the common weed *Chenopodium* – and various teams are investigating the causes, and thus the prevention, of chloroplast decay. However, our approach is to replace the short-lived biological components with more stable synthetic chemicals that can do the same job.

First let us consider the hydrogenase enzyme, the "catalyst" in the hydrogen production process and one of the three key components in a photobiological solar energy system. In the past five years or so our team and a few others have isolated hydrogenase from a number of different bacteria. So far only two hydrogenases have been found to be relatively insensitive to oxygen. These two are from the hydrogen bacterium *Alcaligenes* and the sulphate reducing bacterium *Desulfovibrio*. The hydrogenase from the *Alcaligenes* evolves hydrogen, albeit slowly, from a system containing chloroplast membranes and methyl viologen. This takes place without having to add any oxygen scavengers. The "active centres" of the hydrogenase are clusters of iron and sulphur atoms within it. Researchers have chemically synthesised a number of 4Fe-4S clusters which mimic these biological active centres and we have shown that these iron-sulphur analogue compounds, some with small peptides attached, will replace ferredoxin (but unfortunately not hydrogenase) in the hydrogen producing system. We have only limited knowledge of how the hydrogenase works and how its action differs from that of the ferredoxins. However, we ultimately envisage replacing both these components in the chloroplast system with stable analogue compounds and possibly other compounds such as platinum, or simple particles synthesised from inorganic salts and formaldehyde (formed spontaneously by light).

The inorganic catalyst platinum dioxide can replace hydrogenase in our chloroplast hydrogen-evolving system but only with methyl viologen as the electron carrier rather than ferredoxin and only under anaerobic (oxygen free) conditions. This emphasises three important features that we have to incorporate into any future process. First, there must be a close coupling between the electron carrier and the hydrogen-evolving catalyst. Secondly the efficiency of interaction between the chloroplast membrane and the same electron carriers must be high. Thirdly, the electron carriers must not be easily oxidised. Both ferredoxin and methyl viologen, in their reduced forms, can react with oxygen. Thus they would be little use in a hydrogen production medium that contains oxygen, such as water. In this respect the non-oxidisable biological electron carrier NAD may prove valuable. Chloroplasts reduce NAD, but we still have the problem of producing hydrogen from reduced NAD at reasonable rates.

At present the stability of the isolated chloroplasts limits the duration of hydrogen production, but when photolysis ceases we can revive the system by adding fresh chloroplasts. Despite advances in the stabilisation of chloroplast membranes – for example, by immobilising the enzymes and encapsulating them in alginate gel films – chloroplasts are not likely to be a part of any commercial solar energy conversion system. However, research into chloroplasts' composition and how they work may ultimately provide the basis of a synthetic process. What advances are likely in this area? The water-splitting reactions of the chloroplast membrane (Photosystem II) take place in a manganese–chlorophyll–protein complex; unfortunately we know little about the arrangement of the manganese ions and the mechanism of water splitting in living plants. Despite this lack of knowledge, two research groups seem to be close to splitting water with visible light.

Sir George Porter's group at the Royal Institution, London, has developed manganese porphyrin and phthalocyanine complexes that will generate oxygen and hydroquinone from water. Melvin Calvin's laboratory at the University of California at Berkeley has shown that zinc oxide semiconductor crystals plus dye complexes will effect a similar reaction. Both these groups are trying to mimic natural photosynthesis by incorporating these complexes into liposomes, thus creating charge transfer separation reactions across a membrane – the essence of natural photosynthesis. Calvin's ultimate aim is to have two components; one generating oxygen and a reduced electron carrier by photolysis, the other evolving hydrogen from the reduced carrier in a light-driven reaction. Such a system would be *in vitro* photosynthesis. Although these researchers emphasise that their work is at an early stage, the artificial generation of oxygen and reducing equivalents from water – that is, is Photosystem II type reaction – has been demonstrated, albeit not yet on a sustained truly catalytic basis without using up any substrate.

From a practical point of view, we may have to separate the light-activated step of laboratory photosynthesis that evolves oxygen from the subsequent liberation of hydrogen in a dark reaction – one that does not need light. A one-stage system would generate a mixture of hydrogen and oxygen, and separation and collection over the whole area of any solar collector may be impractical. A two-stage system, however, would consist of a first stage containing the photocatalytic system which generates a non-oxidisable carrier and evolves oxygen. Indeed, collecting the oxygen may prove economically important. The carrier is then circulated to the second stage, and this would consist of the hydrogen-evolving reaction, a dark reaction. The production of hydrogen from the carrier then occurs in a defined volume. After this stage the carrier is recycled back to the first stage of the plant to be reduced in the light reaction, ready to take part in the cycle again.

Photochemists in Russia, France, and Switzerland have shown that an artificial electron donor (in place of water) and light-absorbing pigments (rather than chloroplast membranes) can produce hydrogen. Very recently two groups have claimed that reaction mixtures containing ruthenium compounds and other catalysts could indeed split water with visible light to produce oxygen and in one instance hydrogen as well (see p. 221).

Light-driven hydrogen production is an important step on the way to the ultimate goal of artificial "photosynthesis", but before it can become a part of a solar energy conversion system we have to develop a stable and efficient photosensitiser (the answer here may lie with the ruthenium complexes) and find suitable substrates as electron donors. Another approach is to try to find an artificial photosynthesis process that can "feed on" waste materials, such as sewage sludge. Indeed, Japanese scientists have shown that this can be done using photosynthetic bacteria, but no one has yet demonstrated an artificial process that can do the same thing.

Photobiological hydrogen production from water can be carried out by intact organisms. Both cyanobacteria (blue-green algae) and green algae possess two photosystems analogous to those of green plants, and thus have the ability to split water and evolve oxygen. The blue-green algae first appeared on Earth about 2500 million years ago as a result of depleting organic energy sources – perhaps they can teach us something today. Unfortunately, many problems stand in the way of the development of controlled systems based on these organisms. To begin with we have to achieve and maintain the conditions needed in the cultures that these algae live in; we also have to deal with competing physiological reactions such as nitrogen fixation. A team at the University of California, Berkeley, has reported that continuous cultures of *Anabaena* evolve hydrogen and oxygen for 19 days. However, there are many problems inherent in such whole-cell systems, making it difficult to turn them into solar energy conversion processes.

The net yield of product in a photochemical energy storage reaction is not likely to exceed 13 per cent. If a solar energy system could operate at an overall efficiency of 10 per cent – the amount of incoming solar radiation that ends up as stored product – then the world's total current energy needs could be met by turning over only half a million square kilometres to solar energy collection; only 0.1 per cent of the Earth's surface. This area is about the same as that of Morocco, Thailand, or France, or twice the area of Britain, or one-fifteenth that of Australia or Brazil, or three-quarters of Texas. It would not be necessary to take over agricultural land, and the sea could provide the water. This may seem like science fiction, but it does illustrate the magnitude of the solar resource and the opportunities that lie ahead, even if water-splitting hydrogen production systems yield only a fraction of the world's energy requirements.

Clearly this approach will catch on only if it is economically competitive with other energy sources, unless overriding local considerations force a country to adopt an otherwise uneconomic energy technology. The uncertainties are so great that we cannot calculate the likely cost of a photobiological or photochemical solar water splitting system – such cost estimates can come only after we have a stable system operating for months or years. However, rough calculations show what the potential is, what the limiting factors might be, and what directions the research might usefully take.

In Britain, using a conversion efficiency of 10 per cent, with total solar energy equal to 10^{21} Joules/year, about 100 times the country's total primary energy consumption, we estimate that a tenth of Britain's land area would meet its total energy requirements. In other words, 2 per cent of the land would yield hydrogen equivalent to the country's natural gas consumption; 4 per cent would match our petroleum or coal needs. Britain is not likely to set aside such large areas for "growing" energy, but the area of land needed to substitute solar energy for our natural gas requirements is interesting. Photolytic systems operate at low temperatures, as well as high temperatures, and do not require direct or concentrated sunlight to produce hydrogen gas, a storable energy source.

These crude requirements can give us some clues as to the costs that would be acceptable for a full blown solar-hydrogen process. If we could build a stable system in Britain, where we receive solar energy at an average rate of about 10 million $J/m^2/day$ ($100\ W/m^2$), it would produce about 90 litres of hydrogen/m²/day (about 3 kg of hydrogen/m²/year, with an energy content of about 400 million Joules). This 3 kg of hydrogen would be worth about £1 at current prices.

Thus over a 30-year period the system would earn £30 for each square metre of collector, and therefore the cost of the hardware in the system should be about half this figure. While this is a very crude economic analysis, it does show that the system must be simple, and preferably it should produce hydrogen and oxygen separately, unless, of course, the gases are to be burned immediately.

So what are the short-term prospects for a photobiological/photochemical system that produces hydrogen? That depends upon the progress we make in understanding the science of the process. A crucial breakthrough in current research will come when we attain an efficient and stable system that mimics Photosystem II of chloroplast membranes; that is a process that splits water to evolve oxygen and provides protons and reducing equivalents. The second stage of such a system – the production of hydrogen gas – is well in hand as there are many alternative electron carriers and catalysts to produce hydrogen. However, the sensitivity of many of these components to oxygen may limit their practical use. An important requirement is a substance that can do the same task as hydrogenase; that is, an oxygen stable redox and proton carrier. Just as these iron-sulphur proteins hold the key to all life's reactions, analogues of these ubiquitous proteins may well fulfil crucial roles in hydrogen-producing solar energy conversion systems.

The Dead Sea: Source of a new form of energy?

43

Microbes that capture the Sun

STEVE PRENTIS

15 October, 1981

Purple bacteria that live in the Dead Sea use their outer membranes to
entrap the energy of the Sun. Can we?

The Dead Sea is not totally dead; its inhabitants include a few
unusually hardy microorganisms – and among them is a bacterium
that has already achieved fame as a biological phenomenon,
potentially able to make outstanding contributions to biotechnology. This organism is *Halobacterium halobium* which, in
common with some loosely related species, possesses a mechanism
for harnessing the Sun's energy that has been found nowhere else in
nature. Over the past decade this mechanism has become one of the
most intensively studied biological systems and has provided some
of the most compelling evidence for the "chemiosmotic" theory of
bioenergetics which earned for its creator, British biochemist Peter
Mitchell, the 1978 Nobel prize. In addition, scientists hope that
H. halobium may prove extremely useful – not least for generating
electricity from sunlight and for the desalination of sea-water.

   H. halobium is one of a group of organisms known as halophilic
(salt-loving) bacteria. Salt-loving they certainly are; *H. halobium*
thrives in environments that are up to eight times as salty as the
Atlantic. Indeed, these bacteria quickly disintegrate in a medium
that is "only" three times as salty as the oceans. As it happens, the
ease with which these bacterial cells can be split by putting them in
relatively dilute solutions – a process called osmotic shock – proves
to be very convenient when attempting to make use of the apparatus
they possess for harvesting light in constructing solar-powered
"batteries".

   One of the features that sets *H. halobium* apart from nearly all
other organisms resides in the membrane that envelopes each cell.

H

Up to half of this cell membrane may be covered with purple patches. These patches contain a pigment, bacteriorhodopsin, which can trap the energy in sunlight. If the bacteria can use bacteriorhodopsin in this way, then so perhaps can we.

Bacteriorhodopsin consists of a single protein chain to which is attached a molecule of retinaldehyde. Retinaldehyde — like the light-sensitive pigments of the eye — is a derivative of vitamin A and it is the retinaldehyde part of the bacteriorhodopsin molecule that seems to be "triggered" by light. The much larger protein section of the molecule is folded into seven helical regions, each of which spans the thickness of the cell membrane. When light falls on the purple patches of *H. halobium* it induces chemical and physical changes in the bacteriorhodopsin molecules which result in protons (ions and hydrogen — H^+) being transported from one side of the membrane to the other. Thus bacteriorhodopsin has come to be known as a light-driven proton pump. The whole series of changes that take place in the molecule during this process — the "photocycle" — are completed in about 10 milliseconds, and after each cycle the molecule is again ready to be activated by light and to pump another proton.

The bacteriorhodopsin molecules are arranged in the membrane in groups of three, each group forming part of an hexagonal lattice that can be regarded as a two-dimensional crystal. Evidence has now emerged which shows that the three molecules in each group "co-operate" so that when two or three are working at the same time each increases the efficiency of the others.

In undamaged cells, living bacterium protons are pumped from the inside into the surrounding medium, but the pumping action takes place efficiently even when the cells are destroyed, and only the patches themselves remain. It is this ability that makes purple bacteria such promising candidates for biotechnology.

The act of pumping protons across a membrane, either in the living cell or in the separated membrane, sets up a proton concentration gradient: the protons are most concentrated on what in the intact cell is the outside of the membrane, and less concentrated on the inside. This gradient in protons manifests itself (or can be described) in several ways. The standard measure of acidity, pH, is in fact a measure of proton concentration; that is, the proton concentration gradient is a pH differential. Again, because protons carry a positive electric charge, the concentration gradient also gives rise to a potential difference. These two differentials, the one chemical, the other electrical, are together termed a proton electrochemical gradient. The gradient represents a form of stored energy (and is the prime ingredient in Mitchell's hypothesis).

The living cells of purple bacteria tap this energy store as protons flow back into the organism, providing it with energy for vital processes. The question is: can we develop artificial systems incorporating purple patches to supply our energy needs? What is needed is to arrange purple membranes so that when the Sun shines on them, protons flow from one side to the other – thus creating an electrical potential that can be put to work. No one has yet shown that this can be done economically, but research in several countries is uncovering something of the system's potential and is identifying ways to tackle the problems.

Biochemists, physicists and biophysicists from Israel, the USSR, Britain, Holland, West Germany, Hungary, the US and elsewhere have tried various ways of building the purple patches of *H. halobium* into artificial systems. Even the best systems designed to date fall well short of the efficiencies attained by other methods of harvesting solar energy. Silicon cells, for instance, have exceeded 19 per cent. According to S. Roy Caplan, whose group at the Weizmann Institute in Israel has been in the forefront of research on the purple membrane, and Kehar Singh, the efficiency of present-day *H. halobium* systems must be increased by four orders of magnitude before these systems can become competitive. Laboratory-scale systems have already been constructed and shown to be able to produce electrical currents, and Caplan and Singh have cautiously forecast the "small-scale purple membrane solar energy converters located on the roofs of dispersed or isolated buildings, especially in the Middle East, where the level of radiation is high throughout the year, appear to be a distinct but distant prospect for the fulfilment of domestic energy needs". But at this early stage it is extremely difficult to predict where an investment of time, money, and most importantly, ingenuity might lead.

Purple membrane systems, at least potentially, do have one outstanding advantage over the present "conventional" systems based on silicon, gallium arsenide or calcium sulphide: they are likely to lend themselves to "low technology" production, perhaps needing relatively little capital. Furthermore, although the charge that biological systems are inherently inefficient cannot be lightly dismissed, the concern with efficiencies is not always appropriate. Thus, a system of low efficiency – say 1 per cent – will clearly require 20 times more land (or at least area) than one with a 20 per cent efficiency. But the land surrounding isolated habitations in very sunny areas is often worth little, and it is in precisely such places that purple membrane systems are likely to prove most useful for generating electricity.

The box summarises the technical problems that must be overcome before purple membrane systems are a commercial proposition. As can be seen, the first two problems – breeding and growing suitable *H. halobium*, and separating their purple membranes – are biological. The next three necessary tasks involve the creation of support – a matrix – on which to hold the purple membranes, firmly and correctly orientated. The final problems – inserting electrodes; providing a suitable light source; and linking up the purple membrane solar cell to an electrical outlet – involve conventional electrical engineering. We do know enough to accomplish at least some of these steps – for example, the last stage. But what of the biology?

Growing *H. halobium* should not prove too difficult. The bacteria can probably be grown in simple shallow ponds containing a high concentration of sodium chloride, which is cheap, plus other mineral salts and some nutrients. Unwanted organisms are unlikely to contaminate the growth medium – often a problem when attempting to grow microorganisms on an industrial scale – as very few other species could survive in such an uninviting habitat. Nor

---

How to construct a solar cell based on the purple membrane

1. Grow *Halobacterium halobium* cells which contain as much purple membrane as possible.
2. Collect the cells and separate the purple membrane from the other parts of the cells.
3. Make sturdy supports for the membranes.
4. Fix the purple membranes onto the support so that all the membranes are oriented in the same manner – that is, so that they all pump protons in the same direction.
5. Use the purple membrane/support system to form a barrier between two solutions – the two halves of the electrical cell.
6. Insert electrodes into each compartment.
7. Illuminate the system to power the proton pumps and create a potential difference between the two compartments.
8. Connect arrays of solar cells in series (to increase the voltage) and in parallel (to increase the current).

would the bacteria require additional oxygen – in fact *H. halobium* does not synthesise its purple membranes unless it senses a low external concentration of oxygen. If there is plenty of oxygen the bacteria obtain energy by oxidising foodstuffs – the purple membrane is essentially a standby system when oxygen is scarce and sunlight is plentiful.

In some strains of *H. halobium* half of the cell membrane is covered by purple patches. If new strains can be found that produce even more purple membrane or require fewer nutrients then clearly the economics of this stage can be improved. At present, a 10-litre culture of *H. halobium* has been estimated to yield half a gram of purple membrane, enough to cover about 120 square metres. At an operating efficiency of 1 per cent, Caplan and Singh have estimated the cost of producing this purple membrane at $0.2 per watt. Such efficiency has not yet been achieved – but the cost of materials compares favourably with silicon flat plate arrays at $11, or silicon ribbon at $0.37 per watt.

In all, there seem to be no great problems in obtaining enough *H. halobium* cells: stage 1. Similarly, it is not difficult to separate the purple membrane from the rest of the cell: stage 2. Indeed, the membrane is remarkably sturdy, for its ability to pump protons is little affected by heat, mechanical stress or *p*H. This resilience in the face of violent treatment is more than an interesting quirk of nature; it could prove crucial if purple membranes are to be extracted by fairly unsophisticated, and hence cheap, techniques that may be economically sound in small-scale operations.

Current methods for isolating the purple membrane involve immersing the whole cells in water, causing them to break apart. The cell fragments are centrifuged to separate the various components according to their densities. The purple membrane can then be collected and the proton pump. If halorhodopsin can be purified and handled in the same way as bacteriorhodopsin in the purple membrane, it too might develop into a desalination device. In this case, it would not be necessary to incorporate a H^+/Na^+ antiport. Sodium ions would be removed directly without the intervention of protons.

When *H. halobium* uses its purple membrane to create an electrochemical gradient, its main purpose is to make adenosine triphosphate (ATP) which is able to drive many of the chemical reactions essential for the cell's survival. Just as ATP is essential for this and all other organisms, so supplies of this high-energy compound will be necessary for many biotechnological processes.

ATP has long been referred to as the "energy currency" of living systems; with the expected growth of biotechnological industries the term "currency" promises to take on an altogether more literal meaning. Thus the incentives to develop ways of manufacturing large quantities of ATP cheaply are great. By using light and the purple membrane to drive the natural enzymes that create the high-energy ATP from the "low-energy" adenosine diphosphate (ADP) it may be possible to make ATP on a large scale. The important advantage purple membranes have in this, as in most of its potential applications, is that the system is comparatively simple and does not depend on a whole panoply of other biological systems to perform the function of transducing energy.

The fourth, and probably most speculative, application for the purple membrane is to produce hydrogen and oxygen by splitting water molecules. The photosynthetic apparatus in plant chloroplasts is known to split water and indeed much effort is being invested into making this a practical proposition with a view to using hydrogen as a fuel. Systems incorporating the purple membrane may also be able to do this.

However, another quite different way to exploit *H. halobium* is first to investigate fully how rhodopsin works – the exact physico-chemistry that enables the purple membrane to operate as a proton pump. The approach (which David Hall and his colleagues have discussed in connection with photosynthetic systems) is not to employ biological systems directly, but to learn from them and then imitate them. Lester Packer of the University of California, Berkeley, sums up one of the views on his research on purple membranes: "It will be a long haul to improve materially the existing systems. However, having the basic molecular information on the details of the light energy conversion process could enable us to develop an efficient system patterned after bacteriorhodopsin. For example, knowing the structural features of the molecule, we could in principle synthesise a small segment that might exhibit the activity we desire."

The whole story of the discovery of bacteriorhodopsin and the teasing out of its secrets may become a classic illustration of the technological possibilities that can be thrown up as a result of basic research. When Walther Stoeckenius and Dieter Oesterhelt discovered 10 years ago that bacteriorhodopsin acted as a light-driven proton pump their ideas were seized upon as a means of testing the rival theories of bioenergetics. As the dust begins to settle (although the fierce debate is by no means over) and Mitchell's hypothesis has become firmly ensconced, the practical consequences of a decade of basic research may begin to be exploited.

It is still early days in studies on the biotechnology of purple membranes, and many other forms of solar energy conversion are competing to see which will be the first to claim a significant place in the sun. Although the pressures to allocate priorities are always great, an eclectic approach seems called for – at least until one or more systems have proved themselves technically and economically – for all applications.

44

Can solar scientists catch up with plants?

LIONEL MILGROM

2 February, 1984

Scientists have long envied the ability of green plants to harness solar energy to split water into its component elements. New groups working with semiconductors are trying to mimic the process. But experiments have been plagued by practical difficulties and irreproducible results.

In the field of solar-energy conversion, there is good news and bad. First the bad news. Unless you are a plant, and barring a miracle in semiconductor physics, providing cheap energy by splitting water with light seems further away than ever. The optimism tinged with realism of 1980 has given way to realism tinged with pessimism. But the good news is that some solar scientists think they know enough about the inner workings of their water-splitting systems to begin planning future directions for research.

Our relentless expenditure of energy depends at the moment on non-renewable resources such as oil and coal. These will eventually run out and as nuclear power seems increasingly risky, some researchers have turned their gaze skywards. They are trying to convert the Sun's virtually limitless outpouring of energy into a renewable resource here on Earth.

One of the most seductive routes is to make sunlight split water into hydrogen and oxygen. The reward would be a cheap, clean, and inexhaustible fuel. For when hydrogen in turn is burnt, it produces only energy and water; no pollution and no waste. This is fine in theory. In practice, however, the water-splitting road to a solar future has become blocked by low conversion efficiencies, unrepeatable experiments – and the odd scientific reputation.

There is more than enough energy in sunlight to break the hydrogen—oxygen bonds in molecules of water. The trouble is that water is transparent so, like a boxer punching cottonwool, the energy simply passes straight through and is not absorbed. The problem is how to get enough photons of light energy into the water molecules quickly enough to produce oxygen. If it is too slow, then harmful by-products (peroxides, for instance) may be produced. What makes chemists green with envy is that nature solved this particular problem aeons ago. The solution is called photosynthesis.

Without photosynthesis there could be no life on Earth as we know it. Not only does photosynthesis supply the oxygen we breathe (by splitting water); it also provides the energy gradient that ultimately drives all living things. It does this by trapping the energy contained in sunlight and using it to convert carbon dioxide and water into energy-rich substances called carbohydrates. In the process, water is oxidised to oxygen. The plant uses carbohydrates in its own metabolism, but so do animals that eat the plants. Other animals dine on the plant-eaters and so a chain of energy dependence is created which links us all to the grass beneath our feet.

As carbohydrates are metabolised (converted back to carbon dioxide and water), the energy they contain is liberated and used by an organism to grow, move, reproduce, fart, devise more efficient ways of killing other organisms, write articles on solar energy, and do all the fun things we creatures do. Although the complex molecular dynamics of photosynthesis are becoming increasingly well understood, they are still too sophisticated to be incorporated into practical artificial solar energy devices. But what is a practical device?

Box 1: how plants do it

Photosynthesis is really two photochemical reactions coupled together. The one that evolved first is photosystem I, which converts carbon dioxide photochemically into carbohydrates, the plant's fuel. The second system, called photosystem II, is the photochemical oxidation of water to oxygen. Coupling the two together committed the young Earth to an oxygen atmosphere and the eventual wide diversity of living things.

The all-important first step in photosynthesis is to collect energy from light. In both photosystems, nature uses banks of green chlorophyll molecules and supporting ancillary pigments acting as antennae. These are sensitive to a large part of the solar spectrum that reaches the Earth's surface.

BOX 1 *continued*

ER = *Electron relays (there are about five relay systems) that transfer electrons to photosystem I. Meanwhile, the electrons' energy is tapped and used to make high-energy molecules such as ATP.*

Chl_{II} = *Chlorophyll.*

Chl_I = *A different chlorophyll that is more sensitive to red light. There are several steps between Chl_I and carbohydrate synthesis.*

Chl* = *Chlorophyll excited by light and ready to give up its electron.*

The act of light absorption by chlorophyll in photosystem II excites electrons in special trap molecules at the heart of the antennae, to a higher energy state.

As soon as this happens, the excited electrons pass through a membrane, to the first in a series of relay molecules which convey the electron down an energy hill (for simplicity's sake the diagram shows four relays, though there are probably more). This process gradually taps the energy of the electrons and uses it to make highly energetic compounds. These are later instrumental in making carbohydrates. The electrons finally reach photosystem I, where they are excited again and go on to make more precursors for carbohydrate synthesis.

Meanwhile, electrons lost from the chlorophyll (thanks to photosystem I) are replenished by water molecules nearby. As the water molecules lose their electrons they are oxidised to oxygen, which is liberated by the plant. All the apparatus for photosynthetic water-splitting is close together, moulded in the membrane of the chloroplast. Transfers of electrons and energy can occur extremely quickly and efficiently.

One of the best photosynthesisers is sugar cane, which has an efficiency of around 11 per cent. On a global scale though, photosynthesis is about 1 per cent efficient. This figure represents the percentage of total incident solar radiation that plant photosynthesis converts to energy-yielding compounds, taken over a whole year (some of which time the plant is not growing). And 1 per cent is quite sufficient if you are a member of the vegetable kingdom with only a global ecology to support. But it is not nearly enough to power an industrial society (although it is worth remembering that our vital reserves of coal and oil are the results of past photosynthesis).

Nevertheless, some countries turning biomass energy to industrial tasks. A well-known example is the use of alcohol from sugar cane as a petrol additive in Brazil. Also, sunflower oil makes an excellent fuel for diesel engines. South African scientists have calculated that if farmers were to turn 10 per cent of their land over to sunflowers, the annual crop, when converted to oil, would make them self-sufficient in liquid fuel.

The best silicon solar cells, the ones that power spacecraft, have an efficiency of about 18 per cent. However they achieve this figure only under ideal conditions, for example in a vacuum where there is no atmosphere to absorb sunlight. And silicon cells are still too expensive for large-scale use, and work well only in direct sunlight.

With current technology, silicon cells would need to cover a land area of 16 km^2 to replace the 2000-megawatt power station at Didcot in Oxfordshire. So, over the past decade, chemists have been looking for cheaper and more practical alternatives. The emphasis has rested increasingly on semiconductors and colloids (systems of tiny particles, from about 2 to 50 nanometres in diameter, dispersed in a liquid) as catalysts for the light-induced splitting of water.

With semiconductors, the idea is for light to drive a photoelectrochemical cell. This consists of a semiconductor electrode in contact with an electrolyte (a solution containing positive and negative ions, which conducts electricity). A metal electrode, also in the electrolyte, completes the cell. When light of the right wavelength is shone into the semiconductor, electrons are excited across the band-gap (see Box 2) and a current flows in the cell. The current can be used to split water, if the electrolyte solution is aqueous, and the hydrogen and oxygen can be collected at the separate electrodes. That is the theory. In practice, things have not turned out so well. The right semiconductor has yet to be found.

There are two problems. The best semiconductor materials, such as strontium titanate or titanium dioxide, have large band-gaps and work well only in ultraviolet light, of which there are only tiny quantities at the Earth's surface. Visible light will not do. But the semiconductors that *do* work with visible light, such as cadmium sulphide, gallium phosphide and gallium arsenide, corrode during the course of the photoelectrochemical reaction. This means that the cell dies after only a short time of operation.

Chemists at Oxford University, led by Professor John Goodenough, have been trying for years to produce new semiconductor materials with the right band-gap to absorb visible light and that do not photocorrode. Success has been very limited and, as far as the Oxford group are concerned, the photo-electrolysis of water to hydrogen and oxygen is dead. Elsewhere, however, photo-electrochemical cells, for the production of electricity are being developed.

Back in 1980, solar scientists were predicting cheap and function-

---

Box 2: electrons, molecules and semiconductors

In a simple diatomic molecule such as hydrogen, the electrons occupy a region of space between the atoms called a bonding molecular orbital. Now certain properties of electrons are described, quantum mechanically, in terms of equations with wave-like properties.

Such equations have more than one solution. It turns out that the bonding molecular orbital is not the only possibility for the electrons in the hydrogen molecule. Each can also form an *anti*-bonding orbital, which is at higher energy. This also occupies the space outside the molecule (along the axis joining the two atoms) and does not take part in the binding of the two hydrogen atoms.

In the hydrogen's lowest energetic form (called the ground state), electrons are content to pair up and occupy the bonding orbital. When the molecule is electronically excited, one of the electrons jumps into the anti-bonding orbital, where it will exist momentarily.

Bonding and anti-bonding orbitals come in pairs and the number of each is directly related to how many electrons there are in the molecule: in a tetra-atomic molecule, for example, there are two bonding and two anti-bonding orbitals. In a polyatomic molecule with n atoms there will be $\frac{1}{2}n$ bonding and $\frac{1}{2}n$ anti-bonding orbitals. The energy gap between the individual orbitals decreases as their number increases.

BOX 2 *continued*

Finally, in a solid with an infinite three-dimensional array of atoms, the number of bonding and anti-bonding orbitals becomes so great, and the gaps between them so narrow, that they merge into separate bands. The bonding orbitals form a *valence band*, while the anti-bonding orbitals merge into a *conduction band*.

Electrons in a solid's valence band are localised around individual atoms in the solid lattice. If electrons reach the conduction band, however, they can move through the lattice when an external electric field is applied.

The gap between the two bands (called the band-gap) is the property that distinguishes a solid as a conductor, a semiconductor, or an insulator. When the band-gap is large the solid is an insulator.

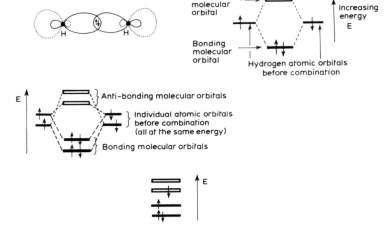

Top: *formation of bonding and anti-bonding molecular orbitals in hydrogen.*
Electrons pairing up in the bonding molecular orbital of hydrogen (H). The dotted line represents the empty anti-bonding orbital. On the right is a state diagram of how electrons pair up in the bonding orbital.
Centre: *formation of bonding and anti-bonding molecular orbitals in a tetra-atomic molecule.*
Bottom: *excitation of a bonding electron into an anti-bonding molecular orbital in a tetra-atomic molecule.*

BOX 2 *continued*

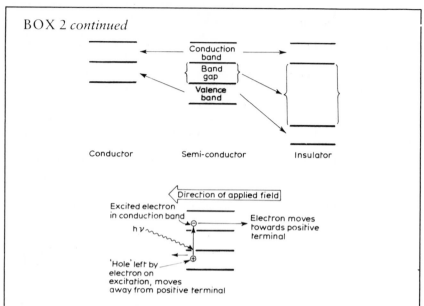

Top: *the band structure of different types of solids.*
Above: *the generation of negative (n-type) and positive (p-type) charge carriers in an undoped semiconductor.*

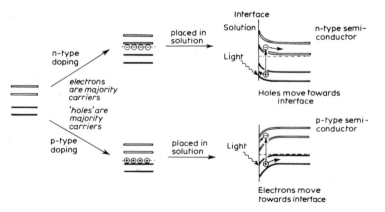

Effects of n- and p-type doping of a semiconductor, and what happens to its band structure in solution.

BOX 2 *continued*

It will not conduct electricity because electrons do not have enough energy to cross the gap.

For metallic conductors, on the other hand, the band-gap is non-existent. This allows electrons to cross easily from the valence band to the conduction band. Here they can roam through the metal lattice, at the whim of any applied external field. Metals may be considered to be atomic lattices awash in a "sea" of conducting electrons.

Semiconductors have a band-gap but, compared with insulators, it is narrow. Thus a small amount of energy from, say, heat or light, will be enough to kick electrons up into the conduction band.

Semiconductors have two modes of conduction. Excited electrons move through the conduction band, in negative or "n-type" conduction. But where the electron has been excited in the valence band it leaves behind a positive charge, called a "hole". This can migrate from atom to atom in the lattice. (The migration is really valence-band electrons hopping into the vacant space.) This constitutes another form of conduction, called positive, or "p-type" conduction.

Under the influence of an external field, the conduction-band electrons move in one direction, while the valence-band electrons move in the opposite.

In a pure semiconductor the number of n- and p-type charge carriers balances out. "Doping" the semiconductor with small amounts of impurities alters the number of charge carriers in favour of n- or p-types. This depends on whether the dopant is electron-deficient (such as the group III elements boron and aluminium) or electron-rich (such as the group V elements phosphorus and arsenic).

When a semiconductor is placed in an electrolyte solution the band structure is distorted or "bent" where it makes contact with the solution and a little way back into the semiconductor. The direction in which the bands are bent (up or down) depends on whether it is n-type doped (up) or p-type doped (down). The band-bending corresponds to the accumulation of one kind of charge carrier at the interface between the semiconductor and solution.

So, if the semiconductor is p-type, with bands bent downward, then electrons excited by light into the conduction band near the interface run downhill toward the solution. Holes in the valence band run uphill, away from the solution, and into the bulk of the semiconductor. The arrival of positive charges at the interface from the solution completes the circuit. A current flows through the cell and into the external circuit as long as light shines. This is the principle of the photoelectrochemical cell.

ing photo-electrochemical cells by the 1990s. The problems were mainly to do with corrosion (because of aqueous electrolytes) and low conversion efficiencies (around 6 per cent). Earlier this year (1983), however, two Californians, Dr Chris Grouet and Dr Nathan Lewis, made a long-lasting cell that was efficient (around 12 per cent) and did not corrode. They used a gallium arsenide–phosphide semiconductor electrode, and replaced the normal aqueous electrolyte with a non-aqueous one based on acetonitrile. The cell has to be hermetically sealed to exclude any trace of water. Also, the electrolyte solution and semiconductors are expensive. The Californian workers are trying to bring the price down, notably by using cheaper amorphous (non-crystalline) silicon as the semiconductor electrode material.

With colloids the situation has not been so clear cut. In fact, colloids could be called the villains of the photochemical piece. Since 1977, laboratories around the world have been experimenting with colloids in potential water-splitting systems. One of the main innovators in this area is Professor Michael Grätzel of the Federal Polytechnic at Lausanne in Switzerland. There is a problem: much of Grätzel's work has been difficult to repeat thanks to the complexity of colloid systems.

Colloids for water-splitting come in two varieties: the hydrogen generators and the oxygen producers. Hydrogen-generating systems consist of a sensitiser, an electron-relay, a colloidal platinum catalyst, and a sacrificial (because it is used up during the reaction) electron donor.

This mixture is suspended in water, and light is shone in. Hydrogen slowly bubbles off and the electron donor is consumed. To keep the system going, the electron donor must be regularly topped up. Other problems are that platinum is expensive and is easily poisoned. The electron-relay most commonly used in the laboratory is methyl viologen. Its other name – paraquat – indicates why it would be dangerous to use on a large scale. Vast lakes of weedkiller scattered about the countryside would be a public health hazard. Also, yields of hydrogen are not large; you could not fill an airship with hydrogen this way.

Grätzel's system was difficult to duplicate at first but now several laboratories have managed it. One of the best hydrogen-producing systems has been perfected at Sir George Porter's Royal Institution in Great Britain. With a zinc metalloporphyrin as the sensitiser, scientists managed to achieve a light-to-hydrogen conversion efficiency of around 60 per cent. Unfortunately, the system's lifetime is only a few hours. Much more work needs to be done before the hydrogen systems will be practical, but, after a fashion, they work. It has left Grätzel, however, with a mixed reputation – solar *wünderkind* or *bête noire* – depending on the colloid preparation.

Meanwhile, theory has begun to catch up with empirical experi-

---

Box 3: the villains of the piece?

Colloids are systems of tiny particles (from 2 to 50 nanometres in diameter) dispersed in a liquid phase. The particles are so small that collisions with the molecules of the liquid keep them in motion. This is known as Brownian movement and stops the particles settling. Colloidal particles will also diffract light, and depending on the particle size different colloids of the same substance will produce a variety of colours.

Colloids consist of a *disperse phase*, which are the tiny particles. They are distributed in a *continuous phase*, which is the liquid. Finally, a stabilising agent, which has an affinity for both the particle and the medium, is required to stop the particles coagulating and precipitating from the continuous phase. This is known as flocculation. Milk and emulsion paints are typical colloids. When milk turns sour, acid is produced, which causes the particles to flocculate: hence the white bits that float on top of tea with sour milk.

The figure on p. 223 shows colloids for generating hydrogen. The sensitiser (S) absorbs the light in the same way as chlorophyll does in photosynthesis. Electrons excited by the light are then passed from the excited sensitiser (S*) to the electron-relay, methyl viologen (MV^{2+}). This is reduced to MV$^+$, which then passes its electron to the colloidal platinum catalyst. The platinum then reduces water to hydrogen and hydroxide ion.

Meanwhile, the oxidised sensitiser (S$^+$) is replenished with electrons by a sacrificial electron donor. Overall, the reaction uses up electron donor and water.

Several things can go wrong with this system. First, the excited sensitiser can easily lose energy before it has had time to pass on its electron to the electron-relay. This happens because all the elements of the system are free to move randomly and collide in solution. Such collisions can rob excited molecules of their energy.

Another problem is that the electron-relay is attacked by the hydrogen that is produced. As the reaction proceeds, the quantity of electron-relay gradually diminishes and the water-splitting grinds to a halt. Also, the hydrogen gradually poisons the catalyst.

BOX 3 *continued*

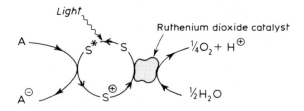

Oxygen generation.

A = *Electron acceptor; a cobalt complex.*
A⁻ = *Reduced electron acceptor.*
S = *Sensitiser (ruthenium(II) trisbipyridyl) doubling as electron-relay.*
S* = *Photoexcited sensitiser.*
S⁺ = *Oxidised sensitiser.*

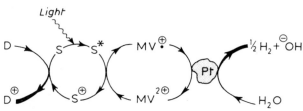

Grätzel-type hydrogen generating system.

D = *Electron donor (triethanolamine, cysteine or ethylene diamine tetra-acetic acid).*
D⁺ = *Oxidised electron donor.*
S = *Sensitiser, ruthenium(II) trisbipyridyl or zinc(II) tetramethylpyridyl, porphyrin, a synthetic cousin of chlorophyll.*
S* = *Photoexcited sensitiser.*
S⁺ = *Oxidised sensitiser.*
MV²⁺ = *Electron-relay, methyl viologen.*

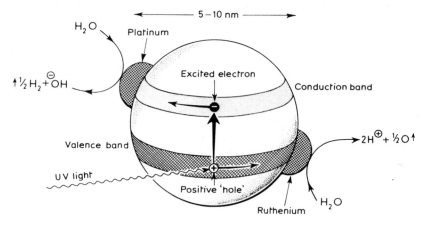

Supposed mode of action of Grätzel's titanium dioxide semi-conductor particles, co-deposited with platinum and ruthenium dioxide. Light pumps electrons into the conduction band, leaving positive "holes" in the valence band. The electrons move to the platinum and reduce water to hydrogen and hydroxide ions. The "holes" move to the ruthenium dioxide and oxidise water to oxygen and protons.

ments. A skilful algebraic blend of electrochemical theory and chemical kinetics (the study of chemical reaction rates) can explain the successes and short-comings of platinum colloids. The man behind this is Professor John Albery of London's Imperial College. He is also one of the founders of the Oxford–Imperial Energy Group, a loose confederation of solar-energised dons at Oxford University and Imperial College.

Albery looks at platinum colloids from two viewpoints – as a system of tiny electrodes, and as a solution of large molecules. The first point of view allows him to treat the platinum colloid particles by electrochemical theory. The second viewpoint means the particles are amenable to the mathematical manipulations of chemical kinetics. The two views are algebraically mixed and the result provides a basis for understanding one of the most serious short-comings of these systems.

As hydrogen is produced, the performance of the system gradually deteriorates. What appears to be happening is that water-splitting is most efficient at sites on the platinum particles that have been

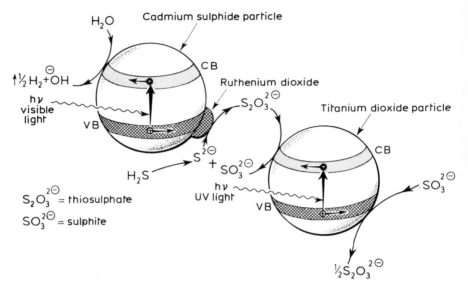

*The cadmium sulphide particle splits water and hydrogen sul-
phide (eventually from oil refineries) using visible light, producing hydrogen and thiosulphate (from sulphide ions). Titanium
dioxide particles protect the cadmium sulphide particles from
oxygen and photocorrosion and convert the thiosulphite to
sulphite and sulphide, with UV light. The system bears certain
resemblances to photosynthesis in that it is a two-stage process.*

oxidised. As hydrogen is produced, it reduces these sites, and
hydrogen production falls off. A further problem is that the electron-
relay, methyl viologen, is reduced by the hydrogen and plays no
further part in the reaction. This is bad news for those wishing to
build practical devices based on platinum colloid particles.

Systems to produce oxygen have fared much worse. Essentially, it
is the same game as the hydrogen generators but in reverse. A
working system has been devised by Professor Jean-Marie Lehn of
Strasbourg University (see diagram). He uses a sacrificial electron
acceptor, which is a cobalt compound. The sensitiser also doubles as
an electron-relay. The catalyst is ruthenium dioxide.

The idea, ultimately, is to put the hydrogen- and oxygen-generating
systems together (without sacrificial donors and acceptors) and
produce a truly cyclic system which photochemically splits water

into hydrogen and oxygen. So far, it has not worked. Which brings us to one of the most controversial photo-catalytic systems of the past 5 years: the semiconductor colloid particle with platinum and ruthenium dioxide deposited on its surface (p. 224). Hydrogen and oxygen are produced simultaneously.

This catalyst is, once again, the brainchild of Michael Grätzel. Brilliant in concept, it has the whole water-splitting reaction occurring on one colloid particle. But it has two main problems. The products of the reaction are not separated, and 2:1 mixtures of hydrogen and oxygen are potentially explosive. The second problem is far more serious. This time, no-one can reproduce Grätzel's results.

The problem is mainly with the oxygen-producing side of the reaction. It does not work and no-one can see photochemically produced oxygen. Grätzel worked originally with colloids made from titanium dioxide or cadmium sulphide. Tiny amounts of platinum and ruthenium dioxide were skilfully deposited onto the particles' surfaces. After the initial success with these systems, other workers *and* Grätzel have failed to see oxygen again. This leaves the twin mysteries of what has happened to the oxygen, and where did it come from in the first place. It is probable that oxygen leaked in from outside the reaction. Any oxygen that is produced also looks as if it combines with the colloid particle's surface. Grätzel now thinks

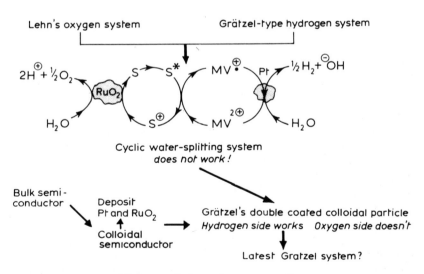

Water-splitting: the story so far.

that putting platinum *and* ruthenium dioxide on the same particle is not such a good idea. Undeterred, however, he has a new system up his sleeve (see diagram, p. 225).

This consists of two different colloid particles. Cadmium sulphide (CdS) particles carry ruthenium dioxide, while titanium dioxide (TiO_2) particles protect them from photocorrosion. The CdS particles split water into hydrogen and hydroxide. They also convert sulphide ions into thiosulphate. The TiO_2 particles then cause both the reduction and the oxidation of the thiosulphate. This system has potential industrial applications because it could deal effectively with the vast amounts of hydrogen sulphide produced during the refinement of oil.

Meanwhile, John Albery has been trying to develop a theoretical approach to semiconductor colloids. He could not understand why Grätzel's particles, when deposited with platinum, did not simply short-circuit (providing a way of electrically connecting the particle's valence and conduction bands). Starting with the bare semiconductor, Albery and his group found that they could be treated once again like large molecules. For instance, when the colloid suspension was made the electrolyte in an electrochemical cell, the particles behaved like huge ions, giving up about 500 electrons to one of the electrodes. This "Hoovering" of electrons, as Albery calls it, can change the colloid particles from n-type to p-type (see Box 2).

When platinum is deposited on the particles, it is as if the colloid particle ceases to be a semiconductor and becomes an insulator. After the particle has been "Hoovered" of electrons, there are none left to act as charge-carriers. The result is that the particle now behaves as a low-band-gap insulator. If that is the case, then why bother with n- or p-type semiconductors? It may be possible to generate hydrogen using colloidal suspensions of cheaper undoped semiconductors coated with small amounts of platinum. But Albery's work suggests another conclusion that could be a useful pointer to future developments in the world of artificial solar-energy conversion. He calculates that each colloid particle receives a photon of light every one-thousandth of a second. By contrast, photosynthesis is so efficient at absorbing photons that four of them (the number required to oxidise two molecules of water to a molecule of water) can be trapped and used in a staggering one-million-millionth of a second! No wonder man-made systems seem so inefficient.

Of course we have a good excuse: research in artificial solar-energy conversion has got into its stride only in the past 10 years. Nature has had thousands of millions of years to get her photo-synthetic act together. So can we learn anything from her?

In man-made systems, all the active constituents of the photo-chemical reaction float freely in solution. The all-important electron-transfer processes that govern the rate of the reaction depend, to a certain extent, upon slow random collisions between these constituents. By contrast, nature ties down all the active constituents of photosynthesis in a membrane. Consequently electron-transfer is *trés rapide* and *trés, trés*, efficient. Perhaps we need literally to take a leaf from nature's book and pack our colloid particles into membranes. As in much of biological research, the more we learn, the more Nature's oddities make sense.

Index